California SURFACE MINING LAW

Publisher's Note

Before you rely on the information in this book, be sure you are aware that some changes in the statutes or case law may have gone into effect since the date of publication. The book, moreover, provides general information about the law. Readers should consult their own attorneys before relying on the representations found herein.

California SURFACE MINING LAW

A Guide to Federal, State, and Local Requirements

Derek P. Cole

SURFACE MINING LAW

A Guide to Federal, State, and Local Requirements

October 2008

Solano Press Books
Post Office Box 773
Point Arena, California 95468

tel 800 931-9373
fax 707 884-4109
email spbooks@solano.com

Cover and book design by
Julie Shell and Pat Shell

Index by Daniel McNaughton,
Chicago, Illinois

ISBN 978-0-923956-73-8

Notice

This book is designed to assist you in understanding environmental law and policy. It is necessarily general in nature and does not discuss all exceptions and variations to general rules. Also, it may not reflect the latest changes in the law. It is not intended as legal advice and should not be relied on to address legal problems. You should always consult an attorney for advice regarding your specific factual situation.

California Surface Mining Law is printed on stock comprised of ten percent post-consumer fiber with soy-based ink formulated for minimally low VOC emissions.

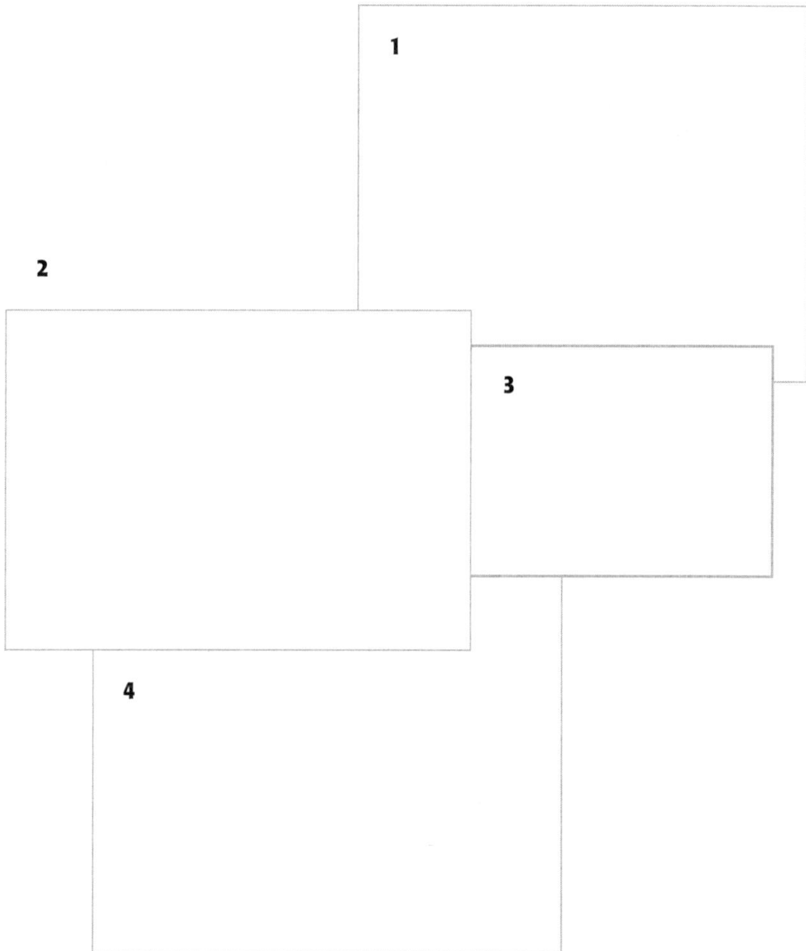

Cover photographs

front

1, 2, 3 Courtesy Granite Rock Company

4 Courtesy The Dutra Group

back left

Courtesy The Dutra Group

Chapters at a Glance

Part One
Understanding Surface Mining

Part Two
Surface Mining and Reclamation Act

Part Three
Local Permitting

Part Four
Environmental Regulation

Contents

Contents

Contents

Part Three
Local Permitting

chapter 6

Enforcement of the Surface Mining and Reclamation Act / 81

chapter 7

City and County Land Use Regulation / 93

Contents

Contents

Part Four
Environmental Regulation

chapter 10

Air Quality / 167

chapter 11

Water Quality / 179

Contents

Photography Credits

Pages 8 and 9, courtesy Granite Rock Company

Page 46

top left, courtesy Larkspur Heritage Committee
top right, courtesy Norman Gilroy
middle left, reprinted with permission from Pacific Aerial Surveys/HJW GeoSpatial
bottom left, courtesy Norman Gilroy

Preface

> *The Legislature hereby finds and declares that the extraction of minerals is essential to the continued economic well-being of the state and to the needs of the society....*
>
> Pub. Res. Code § 2711(a)

With the words that begin SMARA, California's Surface Mining and Reclamation Act, the legislature recognized more than 30 years ago just how dependent we are on mining. Today these words are equally, if not more, true. Nearly everything we use is derived from one mineral or another—highways, roads, houses, hospitals, schools, airports, cars, airplanes, computers, and televisions are just a few examples.

Yet, while mining is so essential to our standard of living, it is hardly a benign activity. Mining is noisy and dusty and can adversely affect its surroundings. Rumbling from heavy trucks and the vibration caused by explosives can be annoying. Ecologically, mining can degrade air and water quality, damage wetlands, and impair sensitive habitat. And, after operations are complete, mined-out land is usually not reusable, safe, and aesthetically pleasing, requiring significant restoration.

Because of these and other impacts, mining, while one of society's most important industries, is also one of its most heavily regulated. A multitude of federal, state, regional, and local laws address its land use and environmental impacts.

This book discusses these subjects and attempts to speak to every audience. Attorneys—whether for mining operators, public agencies, or neighborhood or environmental organizations—will find the book most useful. It is clearly a legal text that covers the laws, regulations, and relevant cases, providing comprehensive citations to guide lawyers in their research and analysis.

But while that audience may be sizeable, the book is not a treatise. Nor was it written strictly for the legal profession. Instead, the author has tried to keep all audiences in mind, attorneys as well as planners, consultants, government officials, professionals, and interested members of the public.

To serve this broad readership, the text does not simply recite the law with strings of citations. Instead, each subject is put in context, describing what the law is and why. And because the surface mining industry may be unfamiliar to some readers, mining-specific concepts are explained. The book offers practical advice on a number of important subjects in an attempt to provide an understanding of the law, as well as guidance on how to comply.

Particular attention has been given to the book's organization. The myriad of regulations affecting surface mining can make the subject seem complex and uninviting. As the adjacent chart shows, as many as twelve authorizations from seven separate federal, state, regional, and local agencies may be required to authorize a new surface mine. To help with what thus may appear a daunting subject, the book is divided into four logical parts, with each chapter covering a discrete set of topics. This organization offers an overview of surface mining while allowing readers looking for specific topics to quickly identify chapters pertinent to their research.

Part One is introductory, directed mainly to readers with minimal understanding of surface mining. Chapter 1, which explains surface mining in its various forms and how it typically operates, may be particularly useful. Chapter 2 describes its uses and applications, touching nearly every facet of our lives and creating an enormous demand with significant land use implications.

Part Two discusses the State Surface Mining and Reclamation Act of 1975, or SMARA. Chapter 3 discusses SMARA's functions and the agencies responsible for enforcement. Chapter 4 addresses the Act's mineral conservation provisions, describing California's scheme for identifying and conserving significant mineral resources. In chapter 5, SMARA's mine reclamation function is discussed, with a description of the mandatory process by which mined-out land must be reclaimed to productive post-mining land use. Chapter 6 covers SMARA's enforcement provisions, focusing on specific enforcement responsibilities the Act imposes on cities and counties and its remedies to ensure compliance.

Part Three addresses city and county regulation. Chapter 7 reviews specific local authorizations a mining operator must obtain, the type of regulations embodied in the authorizations, and the process cities and counties must follow in considering mining applications. How the California Environmental Quality Act, known as CEQA, shapes the permitting process is addressed in chapter 8. Chapter 9 covers the special rules and doctrines afforded to pre-existing or nonconforming surface mines that are not required to comply with local land use regulations.

Part Four covers environmental regulation. Chapter 10 addresses air quality, focusing particularly on how air quality districts regulate particulate matter emissions

Authorizations Potentially Required for a New Surface Mine

Authorization	Responsible Agency	Description	Chapters
Reclamation Plan	City or county	Plan for converting mining operation to post-mining use	3, 5, 6
Financial Assurance	City or county	Financial mechanism to guarantee reclamation	3, 5, 6
General Plan Amendment	City or county	Changes site's general plan designation to allow mining	7
Zoning District Change or Text Amendment	City or county	Changes zoning district or regulations to allow mining	7
Conditional Use Permit	City or county	Regulates mine operations (hours, dust, noise, etc.)	7
Authority to Construct	AQMD	Governs construction of plants and equipment	10
Permit to Operate	AQMD	Regulates PM_{10} (dust) emission	10
Waste Discharge Requirements/ NPDES Permit	RWB	Regulates impacts to surface and groundwater	11
Dredge and Fill ("404") Permit	USCOE	Regulates dredging or filling of wetlands	12
Streambed Alteration Agreement	CDFG	Regulates impact to lakes and streams from debris, sediment	12
Incidental Take Permit/ Habitat Conservation Plan	USFWS	Governs takings of endangered or threatened species	13
Hazardous Materials Business Plan	CUPA	Inventory and emergency plan for handling and storage of hazardous materials	14

AQMD = Air Quality Management District
CDFG = California Department of Fish & Game
CUPA = Certified Unified Program Agency
RWB = Regional Water Quality Control Board
USCOE = United States Army Corps of Engineers
USFWS = United States Fish & Wildlife Service

from surface mines. Chapter 11 discusses mining and water quality through waste discharge requirements and NPDES permits mandated by the federal Clean Water Act. Chapters 12 and 13 address federal and state laws related to wetlands, streams, lakes, and endangered species. And, finally, chapter 14 discusses the handling of hazardous materials that may be used as part of a mining operation.

One final note—from the title, readers will understand that this book only covers regulation of surface mining. Underground gold mines were once commonplace in California, but few active underground operations are left today. Because such mines do not involve surface mining—the removal of materials from openings in the earth's surface—and are not regulated by SMARA, they are not discussed in this book.

Derek Cole
Sacramento, California
September 2008

About the Author

Derek P. Cole is an attorney with Cota Cole & Associates in Roseville, California. His statewide practice includes advising clients on the full range of land use and environmental laws and regulations, as well as representing operators and public agencies in surface mining matters. In addition to his law practice, Mr. Cole has taught legal writing and research at the University of the Pacific McGeorge School of Law, where he received his J.D. in 1999.

Jennifer T. Buckman and **James D. McNairy** are authors of chapter 8, Surface Mining and the California Environmental Quality Act. Ms. Buckman specializes in natural resources law and is a shareholder with Somach Simmons & Dunn in Sacramento. Mr. McNairy is an attorney in the Sacramento office of Seyfarth Shaw LLP, where he is a member of the firm's Litigation Department.

Part One

Understanding Surface Mining

in this chapter...

chapter 1

Surface Mining—Background

Humans began mining 450,000 years ago, making mining, after agriculture, mankind's oldest industry. Initially practiced very crudely, mining involved chipping loose masses of flint into primitive tools and weapons, helping early humans provide for their basic sustenance. As the Stone Age gave way to the Bronze and Iron Ages, however, advancements in methods and technology enabled the uses of mining to multiply, causing mankind's standard of living to correspondingly rise.

Today, little we do and few things we use do not require mining. Mined minerals are used for building, energy, warfare, every type of industry, agriculture, and all types of machinery, computers, and other electronics. Because Californians have become so accustomed to such a high standard of living, it is easy for us to take mining's role in providing these things for granted. But even though many of us may not recognize all of mining's numerous applications, the fact remains that our standard of living depends on its continued existence.

Little we do today does not necessitate mining. Products made from minerals are used for building, energy, warfare, industry, agriculture, machinery, computers, and other electronics.

As important as it is, though, mining is also one of our most heavily-regulated industries. Much of that is because mining can seriously alter the environment and affect our surroundings. As this book will explain beginning in chapter 3, considerable regulations exist at the federal, state, regional, and local level to address the environmental and land use impacts of mining. But before examining those subjects, readers may wish to understand the types of mining operations covered in this book. This chapter provides the first part of that background, defining basic concepts and describing the phases and operations of a typical surface mine. Chapter 2 provides additional background, discussing society's demand for mined materials and the implications of that demand for land use in California.

Definitions and Concepts

Mining involves extraction of any naturally occurring minerals for utilitarian purposes. Minerals are inorganic elements that are classified into three types: metallic, non-metallic (or industrial), and fuel.

To provide an understanding of the various types of mining, it is first necessary to define some basic terms. **Mining** can be defined as the extraction of any naturally-occurring minerals for utilitarian purposes. Howard L. Hartman & Jan M. Mutmansky, *Introductory Mining Engineering* (2d ed. 2002). **Minerals** that are mined are inorganic elements or compounds that have been formed by many years of geologic processes. These minerals may take a variety of forms. They may be metallic, consisting of ferrous metals (*e.g.*, iron and tungsten), base metals (*e.g.*, copper, lead, and zinc), or precious metals (*e.g.*, gold, silver, and platinum). Minerals may also be non-metallic, consisting of sand, gravel, limestone, sulfur, or other substances. Non-metallic minerals are also sometimes referred to as industrial minerals, and there are also fuel minerals, which include oil and gas.

Mining occurs through two basic methods. **Underground mining** involves the removal of minerals from beneath the earth's surface. A typical underground operation would be an underground coal mine. **Surface mining**, the other method, is the removal of minerals from surface openings. An imprecise, although fairly common, term given to surface mining is "strip mining." While strip mining is certainly a form of surface mining, surface mining is also carried out in a variety of other ways, as will be explored in the next section.

SMARA = Surface Mining and Reclamation Act

As is clear from the title, this book covers only the regulation of surface mining. Surface mining is by far the dominant method of mineral production in California. Indeed, the United States Geological Society indicates that only one active underground mine exists in California. United States Geol. Surv., *Mineral Industry Surveys, Active Metal and Industrial Mineral Underground Mines in the United States in 2003*. Surface mining is also the only method of mining regulated by California's Surface Mining and Reclamation Act (SMARA), a principal source of the rules and regulations this book addresses.

Surface Mining Types

- **Open-pit mining**, the most common form of surface mining, is the removal of minerals through benching, a downward progression of excavation in step-like layers
- **Open-cast or strip mining** is the removal of minerals in successive strips
- **Augur mining** involves removal of minerals from the walls or highwalls of mined-out strips with special augur machines
- **Dredging** involves underwater extraction of minerals from a floating machine known as a dredge
- **Solution or heap mining** uses a leaching process to extract minerals from previously excavated piles

Types of Surface Mining

SMARA highlights the various forms surface mining can take, broadly defining "surface mining" as:

> [A]ll, or any part of, the process involved in the mining of minerals on mined lands by removing overburden and mining directly from the mineral deposits, open-pit mining of minerals naturally exposed, mining by the augur method, dredging and quarrying, or surface work incident to an underground mine.
>
> Pub. Res. Code § 2735.

A number of mining methods fall within this definition.

The most common surface mining method is **open-pit mining**, which involves the exploitation of mineral deposits located at or near the earth's surface from surface openings. Typical open-pit operations would include sand and gravel operations and rock quarries. With these operations, excavations are opened usually by the removal of **overburden**, surface layers of uneconomic or marginally economic material that must be removed to expose the underlying mineral deposits. After overburden removal, excavation of the mineral deposits then occurs through **benching**, a downward progression of excavation in step-like layers. Benching, used to maintain safe slopes and to allow equipment access, gives open-pit excavations a bowl-like appearance.

Example of an open-pit surface mine. Excavation of raw material occurs at the bottom of a benched quarry bowl. After being excavated, the material is hauled to the surface and processed in a separate area of the site.

The second most common type of surface mining is **open-cast mining**. As with open-pit mining, this type of mining requires that overburden be removed to expose the deposits to be mined. However, the benching associated with open-pit mining is not utilized as part of this method. Overburden removal instead occurs in strips, giving this method its more recognized name, **strip mining**. As separate strips are mined out, adjacent strips are subsequently opened and the overburden from those strips is cast into the previously-mined strips.

An additional mining method associated with strip mining is the **augur method**. Through this method, strip walls remaining after a strip has been mined, known as **highwalls**, can be bored into to extract additional minerals. Special augur machines are used for this purpose.

Another form of surface mining is **dredging**. Dredging is the underwater extraction of a mineral deposit from a special floating machine known as a dredge. A dredge may use mechanical extraction devices or suction to extract minerals.

Excavated material is often hauled to processing sites with large dump trucks.

Solution or **"heap" mining** is also a form of surface mining. This form extracts the minerals remaining in previously excavated material piles or heaps, such as spoils piles. To allow such extraction, a chemical solution is placed on top of, and allowed to seep into, the piles or heaps, allowing mineral recovery through leaching. The leached minerals are then collected in liners underneath the piles and, from there, transported into solution ponds from which they are processed.

Four Phases of a Surface Mine

Regardless of the method used, a typical surface mining operation will go through four discrete phases during its lifetime: (1) prospecting and exploration, (2) development, (3) exploitation, and (4) reclamation. Each phase is unique, involving distinct activities.

> **Surface Mining Phases**
> - **Prospecting and exploration**
> - **Development**
> - **Exploitation**
> - **Reclamation**

Prospecting and Exploration

The location of mining is constrained by two fundamental considerations: geology and economics. These considerations reflect the reality that

The first phase involves the search for land that contains a mineral resource of a size and grade that an operator can profitably exploit.

no tract of land can be mined unless (1) nature has endowed it with some mineral resource, and (2) that resource is of a size and grade that an operator can profitably exploit. Sites that meet these considerations are anomalies; finding them is like finding a needle in a haystack. Howard L. Hartman and Jan M. Mutmansky, *Introductory Mining Engineering* (2nd ed. 2002). Indeed, because it can be so difficult to find viable mine property, prospective miners may spend considerable time, sometimes even years, looking for a suitable location. However long it takes, this search for a site encompasses the first phase in the life of a surface mine, prospecting and exploration.

Prospecting is the search for potential sites through a review of geologic maps, reports, and literature; review of aerial photography; and other methods. Exploration involves a more refined search of sites identified through prospecting.

The terms **prospecting** and **exploration** technically connote distinct activities, but they are often used interchangeably. Strictly speaking, however, prospecting is the search for potential mine sites through review of geologic maps, reports, and literature; review of aerial photography; and other general search methods. Exploration involves a more refined search to identify and measure a particular location's resources and reserves. **Resources** are the body of demonstrated or inferred minerals contained in a location. **Reserves** are the portions of those resources that are considered economically mineable.

When setting out to determine the extent of a location's resources and reserves, prospective operators may conduct thorough surveying or modeling using a variety of sophisticated methods. Whatever methods or techniques are followed, the goal of exploration is to allow a company to obtain a full understanding of the extent, grade, and characteristics of a particular location's mineral deposit. That analysis forms the basis upon which the company decides whether to mine at that location. If a particular deposit meets the desired criteria, the company will proceed with the next phase of surface mining, development. If the deposit does not, prospecting and exploration will continue.

Development

Development involves the preparation of a mine site for mineral production.

The second phase of surface mining is development, a period spanning from the end of prospecting and exploration to the beginning of mineral production. The focus of this phase is on the preparation of the mine property to support sustained mineral production, a task that is neither simple nor inexpensive.

Development requires a mine operator to make business and property arrangements and obtain all governmental authorizations necessary to permit mining operations.

In the early stages of development, a mining operator must make necessary property and business arrangements. Among other things, the operator will need to secure rights to the mining site, either through obtaining title to the mine property or by leasing it. To fund all of the plant and equipment acquisition and other start-up costs, the mine operator also must obtain financing and may further need to form a new business entity for the mining operation, and make other business-related arrangements.

In addition to attending to those matters, the mine operator must also obtain government permits for many facets of the operation. These pertain principally to the land use entitlements and environmental regulations that are described beginning in chapter 3. Obtaining all these approvals is a lengthy and often daunting task—not to mention one that is fraught with uncertainty. No matter how valuable the mineral deposit to be mined, prospective operators have no guarantee they will be able to secure all necessary authorizations. For reasons that will be explained in the next chapter, this is especially true in California.

The permitting process is also very expensive. Not only can the cost of applications, studies, environmental impact reports, and other required submittals be substantial, but prospective operators will usually incur significant additional costs for professional services. Geologists, biologists, engineers, and other consultants are usually involved in the mine permitting process. So too are attorneys specializing in land use and environmental law.

Once all government authorizations have been obtained—often years after initial applications were filed—considerable work, time, and expense is still needed to physically prepare a mine site for exploitation. Among other things, an operator must construct and install plants and equipment, support facilities, and office buildings and must also construct a road system to support all of the various operations on the property. To open up working areas for excavation, an operator likewise has to clear vegetation and strip and dispose of overburden. Berms and other artificial barriers may further be constructed around the property to provide an aesthetic screen or noise abatement. Overall, this site preparation may require significant earth movement and construction, taking several months or even years.

The development phase also involves physical preparation of a site for mining, which includes clearing, grading, road construction, plant and equipment installation, and related activities.

Exploitation

After the operator has completed the development phase, production of the mineral commodity may begin, bringing the surface mine into its third and usually longest phase, exploitation. Initially, production may be modest, increasing gradually over time. Depending on the size of a site's reserves, production may then occur for many years or even decades.

The third phase of a surface mine is exploitation, the production of the mined commodity.

Exploitation is the most recognized phase of surface mining as it represents the active stage of operations with which many people may be familiar. An explanation of how a typical surface mine operates is provided below. To complete the description of the lifetime of surface mine, however, it is necessary to first discuss the fourth and final phase of surface mining, reclamation.

Reclamation

The final phase in the life of a surface mining operation is reclamation, which begins once all of an operation's reserves have been exploited or once the operator decides to no longer mine. Whereas many mines were historically abandoned following the end of mining, SMARA has since 1975 required that every mining operator reclaim mined property to some productive post-mining use. To comply with that requirement, the operator must remove mining equipment and facilities,

The final stage of a surface mine, reclamation, involves restoration of mined lands to a productive post-mining use.

recontour and revegetate the land to support its subsequent use, and restore disturbed habitat, water courses, drainage patterns, and other aspects of the property disturbed by mining. Even before mine closure, moreover, the operator must prepare for reclamation by obtaining government approval of a detailed reclamation plan and by posting a financial guarantee to ensure that it will complete reclamation. The laws and regulations governing these responsibilities will be discussed in chapters 3, 5, and 6.

How a Typical Surface Mine Operates

As explained above, surface mining can take many forms, the most common of which is open-pit, which includes sand and gravel mines and rock quarries. To provide the reader with some understanding of how a surface mine operates, this mining method will be used to describe—in a very simplified manner—a typical surface mining operation.

To produce a mined commodity, the first process in which a typical surface mine engages is fragmentation, the separation of minerals from rock masses in the earth.

To produce the finished products it sells, the first process a typical surface mine must undertake is **fragmentation**, the process by which rock masses are separated from a mineral deposit. If the deposits are already naturally fragmented, operations can complete fragmentation through the use of ripping blades, dozers, and other forms of mechanical extraction. Because of the mass and hardness of many deposits, however, many operations may need to use explosives as part of the process. When that is the case, operators must periodically engage in blasting.

Where blasting is required, operators must use explosives in portions of their pits or excavations to free up enough raw material for a particular period—a day, a week, a month, or some other interval depending on processing needs and capacities. Each blast involves the drilling of a series of holes in a rectangular or square pattern. After being drilled, each hole is filled with a blasting agent, such as ammonium nitrate. The explosives in the holes are then detonated, although not all at once; each hole is detonated as part of a series, with delays of a few milliseconds between detonations to minimize seismic effects.

A dozer pushes rock down a hill at the Wilson Quarry in Watsonville, where front end loaders load the excavated material for processing.

Once the process is completed, the fragmented materials are hauled to processing areas. A variety of machines may be used for this purpose, including front-end loaders, scrapers, dump trucks, hydraulic shovels, bucket wheel excavators, and drag lines. Conveyor systems, sometimes very long, may also be used to transport the fragmented material.

The next step in the production process involves the conversion of the raw, fragmented material into saleable products. Because the fragmented material may be in large chunks, this usually first requires

comminution, the progressive reduction of the excavated material to smaller, specified sizes. Comminution is accomplished through the use of a variety of crushing plants and equipment. Material may pass through primary, secondary, and even tertiary crushers, depending on the degrees of reduction needed.

Processing involves comminution, the progressive reduction of fragmented material into smaller sizes with crushing equipment.

Excavated material also requires **classification**. In addition to being crushed, the material must often pass through screening equipment to separate it into desired classifications. When screened, material is loaded into a large opening known as a **grizzly** and then sifted through layers of screens made of progressively finer mesh. Screening allows the material to be sorted into the different sizes and types of finished products the operator sells. The classified products are normally then stored in separate stockpiles until needed for delivery.

Some processing operations may further involve **washing**. During processing, material may be washed with water to remove fine and clayish particles, sometimes called **fines**, from the material. This is typical where the excavated material will be used for asphalt, for instance, since the presence of fines inhibits bonding of the petroleum agents used in asphalt to the rock material.

Top. *Excavated hard rock is being processed at the Wilson Quarry in Watsonville.*

Bottom. *Because many mines and quarries have batch plants, cement trucks are a common site at many operations.*

Many mining operations may also produce paved products on site. To allow for such production, these operations will have concrete "batch" plants or asphalt "hot" plants on site for producing Portland cement, which creates white paved surfaces, and asphalt, which creates black paved surfaces.

After processing is completed, the actual sale of an operator's finished products does not occur in the same manner as a traditional consumer transaction. With mining, rather, sales occur on a "job" basis: the mine operator supplies material for specific development or public works projects, usually in large quantities. Independent trucking companies are often retained to haul the material to designated job sites.

To allow for sales in this manner, finished products must be loaded onto trucks for job-site delivery. Separate loading areas may exist for this purpose. Often, the vehicles hauling the finished material are "double-bottom" or two-trailer trucks, which can haul about 25 tons per load. After being loaded, these trucks must pass over a scale before exiting so that the exact weight of their loads may be determined for job-billing purposes. The trucks will also usually pass through wheel washers or rumble grates to prevent mud or dirt caked on tires from tracking out onto public roads. Once they exit the mine site, the trucks then travel to their delivery sites, where the material is delivered to its end users.

After being processed, mined materials are transported to the mining operator's end users, usually by independent truckers.

in this chapter...

chapter 2

California's Mineral Production and Demand

Knowledge of what surface mining is and how it occurs is important to an understanding of laws and regulations discussed in this book. To that end, chapter 1 offers a useful background. Before the relevant laws and regulations can be discussed, however, the reader may benefit from additional discussion of why mining exists, the extent of California's demand for mined products, and land use issues that the state's demand for these materials creates. For reasons described below, these subjects will likely figure prominently in California's future, leading to a heightened importance in the coming years for the laws and regulations detailed beginning in chapter 3.

Importance of Minerals

Society needs an enormous amount of minerals to make the things we need and use. Yet, few realize this. When shown human requirements such as those in Figure 2-1, most people express amazement. As that graphic indicates, every American baby born today will need 3.7 million pounds of minerals, metals, and fuels in his or her lifetime. Of that amount, the largest demand is for crushed stone, sand, and gravel, which equates to more than 1.6 million pounds per person. Those, principally, are the materials used in the **aggregate** that is used for road, highway, and building construction—and that constitute the large majority of minerals produced from surface mining in California.

Nearly everything we use or do involves minerals. An average American born today will need 3.7 million pounds of minerals during his or her lifetime.

While 3.7 million pounds may seem like an unbelievable number for just one person, the figure is much less surprising when considering the many different minerals one person alone uses on a typical day. The average person cannot leave the house without using numerous mineral products. Indeed, the house itself is made up of several types of aggregate, including its driveways and sidewalks and

Figure 2-1

Every American born will need...

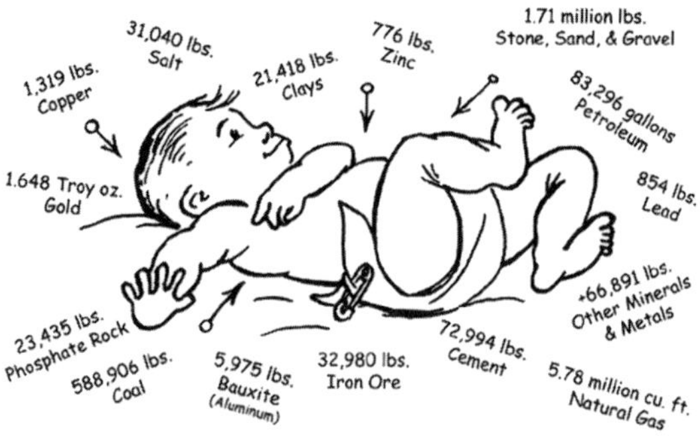

3.7 million pounds of minerals, metals, and fuels in their lifetime

Source: Mineral Information Institute

the foundation on which it sits. Inside, minerals comprise the sinks, countertops, showers, bathtubs, and toilets and are found in its windows and mirrors. Televisions, telephones, and appliances contain circuits, parts, and casings likewise made from minerals. They are also found in the cookware and utensils used for eating.

After leaving home, the average person must use roads and highways that contain asphalt and concrete. Below those paved surfaces are base and sub-base layers made from crushed rock. Additional products made from minerals are the metals and electrical components found in his or her automobile.

The average person will continue to encounter mined products wherever he or she goes. The schools children attend have sidewalks, driveways, foundations, and walls that—like a morning coffee stop, favorite restaurant, or any other building—are made from sand, gravel, and stone. At the office, the average person will use additional mineral-based products, including computers, copying and fax machines, file cabinets, hole punchers, staplers, paperclips, pens, and pencils.

After work, the average person will use roads and highways to return home. Perhaps there will be a stop at the supermarket or a shopping mall to pick up food, household items, other essentials, or luxuries that—if not composed of some mineral products—were all produced and packaged using plants and machines made of mineral-based products that were transported across highways and roads made from aggregate.

In short, there is almost nothing the average person can do and no place he or she can go where minerals are not involved. More importantly, this typical daily scenario is for just one average person in a state with a population of more than 36 million. Consider this same scenario carried out every day by millions of people for several years, and it is not hard to understand why an enormous volume of minerals is needed to support our standard of living.

California's Current Mineral Production

Because of the enormous collective demand for minerals the state's population creates, it should not be surprising that California ranks second among all states in non-fuel mineral production. Susan L. Kohler, Cal. Geol. Surv., *California Non-Fuel Minerals* (2005). From its 820-plus active mines, California produces more than 30 industrial minerals, creating a $3.56 billion industry as of 2005.

Of those minerals, construction sand and gravel was first in production, amounting to $1.269 billion, the highest in the nation. *Id.* Portland cement, which has been used to make white paved surfaces such as sidewalks, ranked second at $1.1 billion, more than every other state except Texas. *Id.* Boron, of which California is the only producer in the nation, ranked third at $483 million, while crushed stone was fourth at $362 million. *Id.* The production of these and all other minerals occurred throughout the state, so that mining affects nearly every region.

As these production numbers make clear, mining is an essential industry in California. And, as the state grows, the industry's importance is likely to grow. This is especially true for the production of construction materials, without which California could not meet the demands of it burgeoning population. Yet, as explained below, the depletion of existing aggregate resources and an unfriendly climate toward mining may threaten the needed growth in the industry, raising concerns about how the state will meet future demand.

Figure 2-2

California 2005 Non-Fuel Mineral Production

Total value $3.56 billion

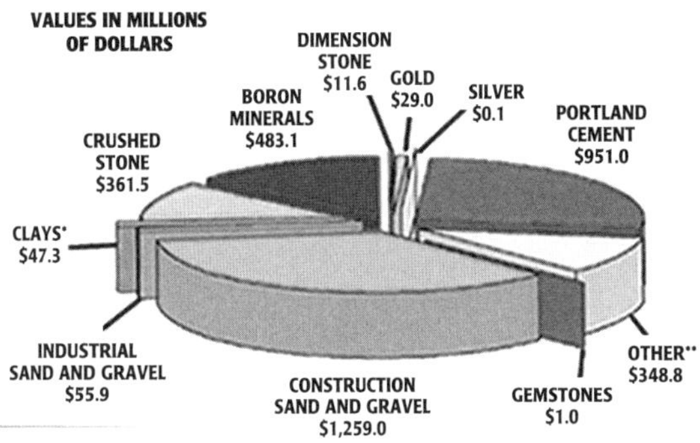

* CLAYS includes: bentonite, common kaolin, and Fuller's earth

** OTHER includes: diatomite, feldspar, gypsum, iron ore, kaolin, lime, magnesium compounds, masonry cement, perlite, pumice and pumicite, pyrophyllite, salt, soda ash, talc, sodium sulfate, and zeolites

Source: California Geological Survey

California's Demand for Aggregate and Future Need for Aggregate Resources

Of all the minerals produced in California, aggregate constitutes the large majority of the state's total production. On average, each resident consumes about seven tons of aggregate annually. Susan L. Kohler, *Aggregate Availability in California, Map Sheet 52*, 16 (Cal. Geol. Surv. 2002). Statewide, that translates to a total demand of 230 to 240 million tons per year. *Id.* Because aggregate is used in every type of road, highway, and building construction, these figures seem only natural for the nation's most populous state.

Every Californian annually consumes about seven tons of aggregate. Aggregate is sand, gravel, and rock used for road, highway, and building construction.

While the aggregate consumption statistics may appear large, though, their size masks the looming supply shortages the state faces. According to a 2002 Geological Survey report, California's existing aggregate mining operations will only be able to meet 56 percent of the 50-year demand for aggregate in urban areas that the state has classified. *Id.*, page 8.

California faces a looming shortage in its supply of aggregate. According to a 2002 study, the state's existing aggregate mining operations will only be able to meet 56 percent of the state's 50-year aggregate demand in urban areas the state has classified.

Yet, even that figure does not tell the full story since the same report found that some urban areas lack sufficient permitted resources to meet even their 10-year aggregate demands. These include Sacramento and Orange Counties, the Western Ventura Region, and San Fernando Valley Region—all areas where substantial population growth is expected. *Id.*, page 10.

Construction aggregate uses

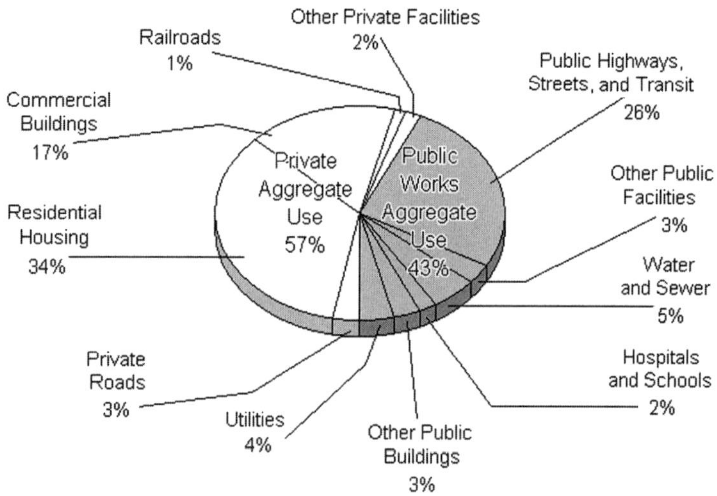

Source: California Geological Survey

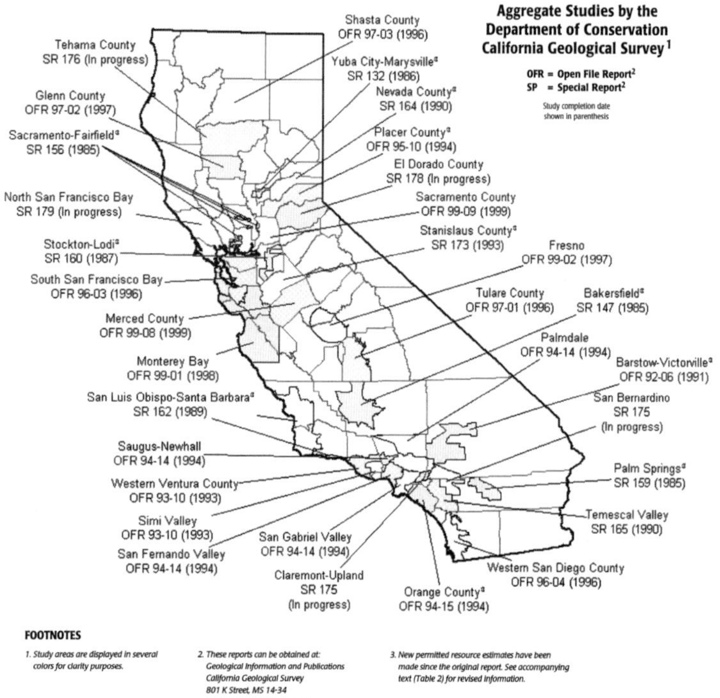

Compounding the looming aggregate shortage California faces is that the state is permitting too few new mines and quarries while at the same time losing existing operations.

Source: California Geological Survey

Other areas and regions of the state fare only somewhat better. According to the 2002 report, the reserves of El Dorado and Placer Counties—both parts of the rapidly-developing greater Sacramento region—are sufficient to meet only 15 percent and 34 percent of their 50-year demands, respectively. *Id.*, page 8. Existing reserves in the North San Francisco Bay Region will only meet 27 percent of its 50-year demand, while existing reserves from the South San Francisco Bay Region will meet 46 percent of demand. *Id.*, page 7. In the Central Valley, the Fresno Region's existing reserves can meet 17 percent of its 50-year aggregates demand, and Stanislaus County's only 11 percent. *Id.*, page 8. In Southern California, the Western San Diego County Region's existing reserves will meet 25 percent of its 50-year demand, Palm Springs Region, 35 percent; the San Bernardino Region, 37 percent; and the Claremont-Upland Region, 50 percent. *Id.*, pages 7–8.

As the problems created by these shortages become more apparent in the coming years, it paradoxically seems likely—if the trends of recent of years are any indication—that new resources will not be permitted to meet the state's burgeoning demand for aggregate. Few new mining operations have been permitted in California in recent years. Indeed, as the 2002 study makes clear, only 6 of 20 areas and regions for which comparisons could be made have reported the permitting of new mineral reserves since studies of these areas began in the 1980s. *Id.*, pages 7–8. Those new reserves also totaled only 109 million tons, less than half the total aggregate consumed in California in 2004. *Id.*

To further compound the problem, California has been losing existing aggregate mining operations as they become depleted or are converted to different land uses. A 1997 report notes that the number of aggregate mines and quarries decreased from 53 to 43 in Southern California including Ventura, Los Angeles, Orange, and San Diego Counties. Russell V. Miller, "Changes in Construction Aggregate

Figure 2-3
Aggregate Availability in California

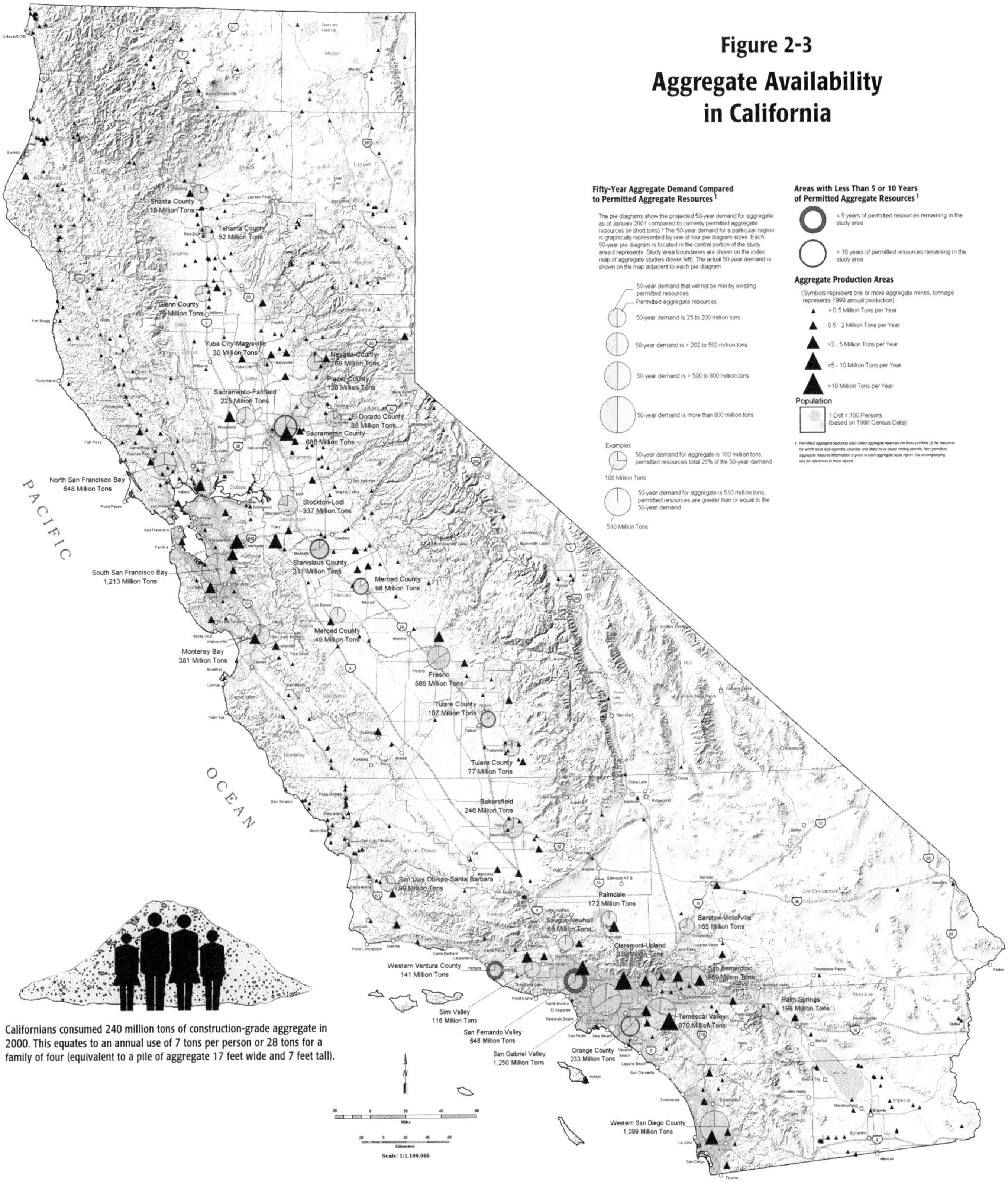

Californians consumed 240 million tons of construction-grade aggregate in 2000. This equates to an annual use of 7 tons per person or 28 tons for a family of four (equivalent to a pile of aggregate 17 feet wide and 7 feet tall).

Source: California Geological Survey

Availability in Major Urban Areas of California Between the Early 1980s and the Early 1990s," *California Geology,* Jan./Feb. 1997, page 4. In the Bay Area, the number of active operations decreased from 24 to 17. *Id.*

Overall, these statistics confirm that the state faces a significant dilemma. To meet the demand for construction sand, gravel, and stone in the next 50 years, California will need more than 12 billion tons of aggregate, which is necessary not only for development of new infrastructure to fuel the state's growth, but also for the repair and rehabilitation of existing highways, roads, houses, hospitals, schools, water treatment and sewage facilities, and other public buildings. But if recent trends continue, it is unlikely that California will grant permits for sufficient resources to meet the ever-increasing demand. This raises the question: how will California meet its long-term need for aggregate? Construction materials have to come from somewhere.

Importing aggregate is not a workable solution for meeting California's long-term demand because it is expensive, its uses are limited, and it presents environmental justice concerns.

As an alternative to permitting new mining operations, two sources for aggregate are available: importing the material from out of state or from recycling. Yet, neither alternative offers a workable solution. At present, California imports only about two million tons of sand and gravel per year, barely enough to make a dent in demand. *California Non-Fuel Minerals 2004, supra*, page 1. The material must also be off-loaded at ports, and from there transported inland by truck. Because the cost of the material doubles every 25 to 35 miles it is trucked, the cost of transporting imported material is prohibitive outside certain limited distances from the ports. *Aggregate Availability in California, supra,* page 12. As a result, importation is not viable to meet most of the state's needs.

Reliance on importation also suffers from two additional problems. First, the material currently being imported consists mostly of common sand and gravel. Other sizes and types of aggregate—of which there are several—are not being imported even though they are equally necessary for many types of development. Second, and perhaps more compelling, is a concern regarding environmental justice. According to one report, the Canadian and Mexican quarries from which California imports sand and gravel have created significant public health and environmental impacts in their surrounding communities. Tom Knudson, "Grounds for Anger," the *Sacramento Bee,* Aug. 17, 2003, A1. Taking advantage of less-demanding environmental regulations, those quarries have sparked outcries from local residents who complain of the destruction of pristine landscapes, health conditions associated with breathing excessive dust, and other problems. Angry to have to bear these burdens for Americans, one Canadian resident remarked, "You're protecting your resources while going after everyone else's. But isn't that one of America's unalienable rights: the right to life, liberty, the pursuit of happiness—and other people's resources?" *Id.*

Like importation, recycling aggregate material does not present a viable solution for meeting California's needs. Although use of the recycling process is growing, recycled aggregate has a limited application due to the low quality of the product produced. Very demanding road and highway specifications limit its use to base rock, which is placed beneath the paved surfaces of roads and highways. *Aggregate Availability in California, supra,* page 9. Recycled aggregate can also be added in small amounts to new mixes of asphaltic concrete (*Changes in Construction Aggregate Availability, supra,* page 5), but does not have other uses.

Recycling is also an unworkable solution for meeting California's long-term aggregate demand as the uses of recycled material are limited.

Because importing and recycling do not present viable alternatives for meeting California's needs, California's only solution to the looming shortages is to permit new aggregate resources. Yet, as will next be discussed, the present climate toward the permitting of new mining operations indicates that achievement of that solution may be unlikely.

Climate for Surface Mining in California

No one can dispute that California will need new aggregate sources to meet demand. Even if the state were to stop growing, substantial amounts of construction materials would be needed just to maintain and repair existing roads, development, and infrastructure. Because of the shortcomings in both importing and recycling, the permitting of new resources appears to be California's only solution for meeting this need.

At the same time, very few people want to live near a surface mine. Hearing of a proposed new project nearby, most citizens do not have a positive reaction. It's the prevailing view that an industrial mining operation cannot be a good neighbor. And fearful of potential harmful effects, residents may be tempted to ask: couldn't the operation be located elsewhere—somewhere away from so many people?

By a cruel hoax of nature, the answer is no. Geologic factors, not demographics, determine where minerals can be mined. Stated otherwise, "Aggregate occurs where Mother Nature put it, not necessarily where we need it. . . ." William H. Langer, U.S. Geol. Surv., *Managing and Protecting Aggregate Resources, Open File Report* 02-415, 1 (2002). And just because a particular location contains aggregate does not mean that those deposits can feasibly be mined. The deposits must meet specific quality standards for their intended uses, which further restricts the locations where these resources can be developed. *Id.*

Geology determines where minerals may be mined. So too does the quality of a mineral deposit.

Economic factors also play a significant role in determining where aggregate can be mined. This is primarily because significant costs are associated with transporting the material. As mentioned above, the cost of aggregate doubles every 25 to 35 miles it is transported. Russell V. Miller, "Changes in Construction Aggregate Availability in Major Urban Areas Between the Early 1980s and Early 1990s," *California Geology,* Jan./Feb. 1997, page 12. To avoid excessive cost, therefore, the production must occur close to the areas that will consume the products. Hauling aggregate from isolated locations hundreds of miles away is not practical.

Because the cost of aggregate doubles every 25 to 30 miles it is transported, aggregate mining must occur close to the areas that will consume the material.

The more truck miles required to haul aggregate, the greater the air quality impacts from transporting the material.

Nor would hauling from long distances be beneficial from an environmental standpoint. Increasing distances trucks must travel to transport aggregate increases diesel emissions and traffic congestion. Since many urban areas in California are already considered "non-attainment" air-quality zones, the additional hauling distances would only exacerbate adverse air-quality impacts. *See* chapter 10, National Ambient Air Quality Standards (discussing areas considered "non-attainment" areas under federal air quality regulations). Perhaps counterintuitively, then, the closer the mining operation to the urban uses it serves, the less the overall impact may be.

Yet, while mining may need to occur near urban uses, the permitting of new operations has become almost a Herculean task. Intense opposition to proposed mining operations is now virtually a given. Overflow crowds often pack public hearings concerning proposed new or expanded mines, with only a few usually attending in favor of the operations. California's notice and public hearing requirements and environmental laws make the approval process a lengthy one—sometimes lasting for years. Couple that with the almost inevitable legal challenges that follow mining project approvals, and it is easy to understand why the permitting of new operations is challenging.

Permitting difficulties have likely had an effect on the filing of applications by the industry. According to one mining industry survey, California ranks near the bottom of the states and countries surveyed in the overall attractiveness of its policies. The Fraser Institute, *Annual Survey of Mining Companies 2004/2005*, 6. Nearly 75 percent of the industry respondents viewed "uncertainty concerning the administration, interpretation, and enforcement of existing regulations" in California as either a strong deterrent to investment or a reason for not pursing exploration there. *Id.*, page 25. The same percentage viewed California's environmental regulations as either a strong deterrent to investment or a reason for not pursuing exploration in the state. *Id.*, page 27. These views, of course, are only those of the mining industry, not necessarily those of California residents. But whether perceived or real, the industry's views are most likely an outgrowth of permitting difficulties it has faced.

The industry's history of abandoned mines has created a poor public perception of mining. And, because many do not recognize the need for mining, it's easy for the public to prefer that land be used for other purposes.

Why has it been so difficult to obtain a permit for a new mine or quarry? One reason is the legacy created by numerous abandoned mines dotting the landscape of California and many other states, particularly in the West. M. Sengupta, *Environmental Impacts of Mining* (1993). Although a relic of the past, abandoned mines have left environmental and physical problems that linger today, creating a low public perception of the industry. Add to that a general lack of awareness of how much society depends on mining. *Managing and Protecting Aggregate Resources, supra*, page 2. Unaware of the true extent of their own personal mineral consumption, many people do not recognize the need for mining, making it easy to prefer that a potential site be used for another, more socially acceptable land use.

NIMBY = Not in My Backyard

Because of these factors, one can't help but observe that proposed mining projects may have fallen victim to the "Not in My Backyard," or "NIMBY," syndrome.

In California, the planning process is often dominated by the "politics of proximity," in which land use decisions respond to the interests of the people and organizations with the most direct stake in the outcome of a land use decision. William Fulton and Paul Shigley, *Guide to California Planning* (3d ed. 2005). While a proposed project may provide an important long-term source of minerals benefiting an entire region, the only people likely to express an opinion about the mining operation are those who have to live nearby. And often, that opinion is overwhelmingly against the proposed operations. Many times, no amount of restrictive conditions or mitigation could be imposed on the operations that could lessen the opposition. For whatever reasons, rational or otherwise, some people simply cannot accept ever having to live next to a mine or quarry.

In sum, however mining is perceived, it is clear that California will have to address its long-term need for aggregate somehow. It will not be possible to avoid addressing the tension between the state's need for aggregate and public opposition to mining, particularly as aggregate shortages become more acute.

NIMBYism

Not all opposition to mining operations can be dismissed simply as NIMBYism. Mining is an intensive industrial use. Even with mitigation, its impacts can sometimes extend beyond its property boundaries. The mining industry's hours, which often require early morning and late night operations to meet demand, also do not mesh well with residents who would prefer not to be exposed to industrial noises at those times.

Mining can also degrade air and water quality, affect sensitive habitats or wetlands, and cause losses to other resources that may not easily be replaced. Just because a site may contain an important deposit of minerals, therefore, does not automatically mean that it can or should be mined. ■

Part Two

Surface Mining and Reclamation Act

in this chapter...

chapter 3

Introduction to the Surface Mining and Reclamation Act

A principal source of the laws and regulations governing surface mining operations is the State Surface Mining and Reclamation Action of 1975, or SMARA, as the Act is commonly called. Because of mining's uniqueness, SMARA imposes a number of duties, both on the industry and the cities and counties that regulate the operations, that do not exist for other land uses. This chapter will introduce SMARA, providing an overview of its functions and the agencies responsible for enforcement. Following the introduction, chapters 4 through 6 will discuss the Act's provisions in detail.

SMARA = Surface Mining and Reclamation Act

SMARA's Two Functions

SMARA serves two functions. It promotes the conservation of mineral resources, explained further in chapter 4, requiring that cities and counties recognize and protect important mineral resources within their jurisdictions. And the Act promotes the productive post-mining use of mined land, described in chapter 5, by mandating that every site undergo a restorative process known as **reclamation** following conclusion of the operations. While each function is explored in detail in a separate chapter, a brief overview is provided below.

SMARA serves two functions, mineral conservation and mined land reclamation.

Mineral Conservation

The majority of SMARA's provisions deal with the reclamation of mined land. Yet, while the bulk of its provisions may be devoted to that function, viewing the Act as purely a reclamation statute is a mistake. The major impetus for the enactment of SMARA was the rapid depletion of resources due to incompatible urban development on and near mineral lands. Recognizing that additional future loss of

The major impetus for SMARA's enactment was the rapid depletion of mineral resource lands.

mineral resources would be irreversible and contrary to California's well-being, the legislature created mechanisms in the Act to identify important mineral resources requiring protection by cities and counties.

With SMARA's dual state-local approach, the state generates information about important mineral resources that local agencies must then incorporate into general plan conservation policies and apply to land use decisions affecting those resources.

SMARA's conservation provisions employ a dual state-local approach to mineral resource conservation. Utilizing the technical expertise of the California Geological Survey, the Act requires that the state's Department of Conservation classify those areas of California containing "significant mineral deposits...subject to urban expansion or other irreversible land uses which would preclude mineral extraction." Pub. Res. Code § 2761(a), (b)(2). Once identified, SMARA then provides an additional process in which the State Mining and Geology board may designate classified areas as having either significant statewide or regional mineral resources. *Id.* § 2790.

A principal function of SMARA's classification and designation is to inform cities and counties—the agencies that have land use authority over mineral resource lands—of the existence and importance of these resources within their jurisdictions. But the consequences of classification and designation are more than informational. Once a site is recognized as an important mineral resource, cities and counties must conserve the resource and protect it from urban encroachment and incompatible development. SMARA further requires that local agencies adopt mineral resource management policies in their general plans to protect the resources, and mandates that cities and counties make special findings whenever a land use application either involves the resources or threatens their potential extraction.

Mined Land Reclamation

Abandoned mines are an unfortunate relic of mining's past. Before the enactment of SMARA in 1976, many mining operators had abandoned the sites following the conclusion of mining, leaving the properties in unsafe, unsightly, and unusable conditions. Through its enactment, SMARA sought to end this practice. Since it went into effect, the Act has required that every mining operator restore the mine site to a productive land use following the conclusion of the operations. The process through which this occurs is called **reclamation,** which generally requires that operators prepare their land for its post-mining or **end use** by removing mining equipment and facilities, reshaping and revegetating the mined lands, and remediating any adverse environmental conditions that remain.

SMARA requires that mine operators restore mined properties to productive post-mining use, a process known as reclamation.

Reclamation is intended to minimize adverse environmental effects, protect public health and safety, and provide for beneficial post-mining use of mined land.

SMARA makes clear the reasons for and benefits of reclamation in a number of its provisions. In stating the Act's findings, the legislature recognized that "the reclamation of mined lands is necessary to prevent or minimize adverse effects on the environment and to protect the public health and safety." Pub. Res. Code § 2711(a). It further found that mandatory reclamation "will provide for the protection and subsequent beneficial use of the mined and reclaimed land." *Id.* § 2711(b). Based on these findings, the legislature declared that its intent in adopting SMARA was to ensure that "[a]diverse environmental effects are prevented or minimized and that mined lands are reclaimed to a usable condition which is readily adaptable for alternative land uses." *Id.* § 2712(a). It additionally expressed an intent that "[r]esidual hazards to the public health and safety are eliminated." *Id.* § 2712(c).

To ensure that the sites are reclaimed, SMARA requires that every mine operator prepare and obtain city or county approval of a reclamation plan, a comprehensive document identifying the operation's end use and specifying steps an operator must take to prepare the site for that use. Mine operators must post a financial assurance, such as a bond or letter of credit, to ensure that their reclamation obligation will be fulfilled. SMARA also imposes other duties to require compliance, including the obligation that an operator submit an annual report and that cities and counties annually inspect mining operations. These duties will be discussed further in chapters 5 and 6.

SMARA requires that mine operators have an approved reclamation plan and post financial assurances to guarantee fulfillment of reclamation obligations.

SMARA's "Permit" Requirement

In addition to the myriad of provisions governing SMARA's conservation and reclamation functions, the Act also tersely provides that every mining operator obtain a **permit**. Pub. Res. Code § 2770(a). SMARA's inclusion of this requirement is curious given how little attention and clarification the Act provides about what it means. SMARA simply states that "no person shall conduct surface mining operations unless a permit is obtained from... the [city or county] for the operation pursuant to this article." *Id.* It does not indicate the type of permit required or the kind of matters a permit should address. Aside from the language quoted, the Act says nothing more of the permit than to offer an unilluminating definition, namely, "any authorization from, or approval by, a lead agency, the absence of which would preclude surface mining operations." *Id.* § 2732.5. To make matters worse, the legislative history of the Act sheds no light on the subject, as there is scant mention of the permit requirement in the legislative record.

SMARA requires that every mining operation obtain a permit from its governing city or county.

Absent a clear indication of the permit's purpose, one can only guess at its function. It is unlikely that the legislature intended that the requirement for a permit have great significance, given the SMARA provisions discussed below. The authors of the Act were very careful to make clear that SMARA did not infringe upon the zoning and police powers of cities and counties. Aside from requiring that they protect lands classified as or designated significant mineral resources and from requiring the reclamation of mining operations, SMARA does not affect how cities and counties regulate mining from the standpoint of land use.

As chapter 7 explains, the likely effect of the requirement is to make mining a **conditional use**, one that can only be authorized upon the issuance of a **conditional use permit**. An additional implication is that any non-permitted mining operation existing at the time of SMARA's adoption likely became a nonconforming use, subjecting it to the rules and doctrines discussed in chapter 9, which deals with nonconforming uses. Because these subjects involve local permitting matters, they are addressed later in this book.

What SMARA Does Not Do

Although SMARA's reach is broad, it is important to understand what the Act does not do, particularly since mining is already subject to a number of other land use and environmental regulations. Most notably, the Act does not preempt or usurp traditional city and county regulatory powers. To address the concerns that local agencies had expressed before the Act's adoption, SMARA's author added language to make clear that it does not restrict:

> [T]he power of any city or county to regulate the use of buildings, structures and land as between industry, business, residences, open space (including agriculture, recreation, the enjoyment of scenic beauty, and the use of natural resources), and other purposes. *Id.* § 2715(f)

SMARA expressly does not limit cities' and counties' zoning powers.

With this language, the Act expressly provides that SMARA does not limit zoning powers. Subject only to the special requirements for significant mineral resources, discussed in the next chapter, cities and counties may zone and regulate mining uses within their jurisdiction in accordance with their normal police powers. The Act comes into play after they choose to allow mining at particular sites and, even then, regulates only how the sites must be reclaimed, not the conditions under which they operate. *See id.* § 2757 (indicating that state policy for reclamation of mined lands shall not pertain to matters "which are solely of local concern... such as, but not limited to, hours of operation, noise, dust, fencing, and purely aesthetic considerations").

SMARA also does not restrict the power of any state agency, city or county, or person to abate nuisances created by mining operations. *Id.* § 2715(a), (b), (d). Nothing about the Act therefore prohibits any public entity or private party from obtaining relief from mining-related nuisances such as excessive dust, noise, light, or other adverse conditions.

Responsible Agencies

SMARA employs a dual state-local approach in its enforcement scheme. While cities and counties, referred to as **lead agencies,** have the principal role in enforcing the Act, various agencies of the California Department of Conservation also play important roles within the overall scheme. The sections below describe each responsible agency and its duties.

California Department of Conservation

SMGB = State Mining and Geology Board

Many agencies within the Department of Conservation play critical roles in administering and enforcing SMARA. These include the State Mining and Geology Board

(SMGB), the State Geologist, the California Geological Survey and the Office of Mine Reclamation (OMR).

OMR = **Office of Mine Reclamation**

State Mining and Geology Board. The State Mining and Geology Board is a nine-member body organized under the Department of Conservation, appointed by the Governor, and confirmed by the Senate. Composed of individuals with scientific training and background, the SMGB acts principally as SMARA's policy-making body. Since its adoption in 1975, SMARA has promulgated numerous regulations to interpret and implement the Act's mineral resource conservation and reclamation provisions.

In addition to its policy-making role, the SMGB oversees city and county enforcement of SMARA and may, as explained in chapter 6, take certain actions against agencies whose enforcement are deemed deficient, even assuming their enforcement powers. In some situations, the SMGB may take action against mining operations, either administratively or by civil action. The SMGB can also hear appeals of permitting, reclamation plan, and financial assurance decisions by lead agencies.

The SMGB, the chief policy-making body for SMARA, may take actions to enforce SMARA in certain situations.

State Geologist and California Geological Survey. The State Geologist and his or her staff within the California Geological Survey have essential roles in carrying out SMARA's mineral conservation function. Under SMARA, the State Geologist is charged with the significant, though enormous, task of classifying all of California's mineral resources. Given the state's size, this is not a task that can be done easily or quickly. For several years, therefore, the California Geological Survey has been administering the Mineral Land Conservation Project, which takes an inventory of the state's non-fuel mineral resources. This ongoing project identifies the location and extent of important resources cities and counties need to accord them the protection SMARA requires.

The State Geologist and California Geological Survey generate mineral resource information for carrying out SMARA's conservation function.

Office of Mine Reclamation. While the California Geological Survey assists in carrying out SMARA's classification of minerals, the Office of Mine Reclamation assists with another major requirement. Through its Reclamation Unit, the OMR provides cities and counties with technical assistance to help evaluate reclamation plans and financial assurances. A separate OMR Reporting and Compliance Unit monitors operator compliance with reclamation plans and other duties, including submission of annual mining reports.

OMR provides technical assistance in reviewing reclamation plans and monitoring operator compliance.

OMR's assistance to cities and counties is integral to SMARA's reclamation function. Because mining is a unique land use, many planners and local officials lack experience in dealing with these matters, particularly mine reclamation. The OMR staff is trained in the relevant scientific disciplines and can help local agencies handle mining-related matters with which they may lack experience.

Lead Agencies

At the local level, the cities and counties responsible for administering and enforcing SMARA are called **lead agencies**. Under SMARA, lead agencies are principally

SMARA describes cities and counties as lead agencies. These agencies approve reclamation plans and financial assurances and also have a number of enforcement responsibilities.

responsible for enforcement of the Act. When lands within their jurisdiction are classified or designated as important mineral resources, these agencies must protect the resources. And, as explained in chapter 6, lead agencies must also approve reclamation plans and financial assurances, annually inspect mining operations to ensure compliance with SMARA, and take appropriate enforcement actions whenever an operator is not complying with the Act.

A carrot-and-stick analogy is appropriate in describing the relationship between lead agencies and the Department of Conservation. The carrot is that the department provides the scientific expertise and technical assistance lead agencies need to carry out their duties to conserve mineral resources and to ensure that mine sites are properly reclaimed. However, when lead agencies fail to fulfill their SMARA duties, the stick is that SMGB can take over some or all of the SMARA enforcement powers and, in some instances, can even approve reclamation plans that otherwise would be approved by the lead agencies.

While lead agencies are principally responsible for enforcing the Act, the state can take over some or all of their authorities when enforcement is deficient.

chapter 4

Identification and Conservation of Significant Mineral Resources

Chapter 2 covers the importance of surface mining to California's future and the looming shortages of construction materials the state faces. In the coming years, as the state grapples with the land use and environmental implications of expanding demand, mineral conservation is likely to garner significant attention.

Fortunately, California does not lack the regulatory means for ensuring an adequate mineral supply. This chapter describes how significant mineral resources are classified and designated under state law, as well as the specific responsibilities of cities and counties to protect these resources. To put these provisions in context, this chapter also discusses the reasons for their existence, stemming from a justifiable concern beginning nearly four decades ago that urbanization was too quickly consuming and encroaching on mineral resources the state would need to meet future demand.

History of California's Efforts to Conserve Mineral Resources

California, as the nation's largest producer of non-fuel minerals, devotes more land to mining than any other state, a status not achieved overnight. While integral to California's founding, mining has been important throughout the state's history. By the early 1970s, when legislative efforts to conserve mineral resources had begun in earnest, California already ranked fifth among the states in total land area devoted to mining. *Letter to Governor Brown from Senator Nejedly* (Sept. 15, 1975). Products produced from mining were then valued, in 1970s-era dollars, at between $750 million and $1 billion annually. Assembly Committee on Resources and Land Use, Analysis of SB 756, page 1 (Aug. 5, 1975).

By the 1970s, California ranked fifth among states in land area devoted to mining. Mining was also then a significant part of the state's economy.

In the years before SMARA's adoption, public officials were concerned about the impact of urbanization on mineral resources needed to fuel growth. They were also concerned that local agencies were making land use decisions that ignored regional and statewide needs for mineral resources.

Already vital to the state's economy when SMARA was enacted, mining was subject to the same threats from urbanization and encroachment as it is today. These looming threats troubled forward-thinking public officials who recognized that, while mineral resources would be needed to fuel California's growth, that same growth—if unchecked—might swallow up land needed to develop those resources. Lacking information about the nature, quantity, and location of significant resources within their jurisdictions, it was all too easy for local planners and decisionmakers to make land use decisions guided purely by parochial interests, ignoring the needs of the region and the state. Indeed, according to a report written a few years after the adoption of SMARA:

> Aggregate mineral resources that would have been adequate to meet the needs of [the San Francisco, Los Angeles, and San Diego metropolitan regions] for thousands of years have been urbanized to the extent that the remaining available supplies in some regions are now sufficient for only a few decades....
>
> Ralph C. Loyd, *Mineral Land Classification in California*, California Geology, Mar. 1982, pages 49–50

SMGB = State Mining and Geology Board

After a critical mass of concern developed over California's ability to provide for its long-term mineral needs, the state in the late 1960s began to explore the adoption of a statewide mining policy. In 1967, a Senate committee requested a study of "the subject of uniform controls and standards for strip mining...." Report of the Committee on Surface Mining, at App. A (Oct. 27, 1970). That request was then referred to the State Mining and Geology Board (SMGB), which appointed a seven-member committee to explore the "problems presented by surface mining." *Id.* at App. B. However, the committee was asked to do more than address the problems created by mining; it was charged with exploring the effect of urbanization and encroachment on mining uses, a problem the Secretary of the Resources Agency summed up as follows:

To address concerns about the state's ability to meet its long-term mineral needs, the SMGB appointed a special committee to explore urbanization's effects on mining uses.

> Recognizing that increase in population inevitably means increasing urbanization, and recognizing too that increase in our standard of living inevitably places increasing demands on the mineral industry, it is to be expected that in some places these two developments are likely to find themselves on a "collision course."
>
> *Id.* at App. C

Appointed by the SMGB and chaired by an economics professor, the Surface Mining Committee included representatives from both the mining industry and state and local government officials. After several hearings, visits to mine sites, and extensive testimony, the committee issued a report in 1970. Among its principal conclusions was that:

A principal conclusion of the Surface Mining Committee was that local agencies made land use decisions without considering the importance of mineral resources.

> Local land use planning and control in some instances has been carried out by local planners and officials without recognition of the existence or value of mineral-resource sites or of the importance of such minerals to the State, to nearby regions, or even to the locality itself. This disregard can be explained in part, by the paucity of

information available to local planners and officials on mineral resource deposits, on the market demands for them and on conservation plans and policies.

Id., page 20

After identifying the problem, the report highlighted its significance, concluding:

The committee considered the deficiency of state controls and efforts to protect mineral resources to be serious. It recommended that the state implement policies emphasizing mineral resource conservation.

In view of the importance of sustained mineral production to the general welfare of the state, and the increasing conflicting surroundings, particularly, urban surface-mining operations and proposals, the deficiency of State efforts and State controls in this area would appear to be a serious one.

Id.

Fortunately, the report did not simply stop at identifying the problem, which by then had become obvious to most observers. Instead it offered practical recommendations. These included proposals for a state-commissioned inventory of California-wide mineral deposits and adoption of a state policy to promote their conservation. The report recommended that:

[T]here be devised a fundamental State policy directed toward the conservation of mineral resources, the conduct of surface mining, and the reclamation of mined land. The policy should represent the full range of public interest in mineral availability and land use. The policy should be made statutory law and be reflected in the relevant plans and programs of State and local agencies.

Id., page 22 (emphasis in original)

As a means of implementing these recommendations, the report advocated that local agencies include mineral information and policies in their general plans. More specifically, it suggested that agencies utilize the general plan process to:

The committee advocated that local agencies be required to include mineral resource information and conservation policies in their general plans.

(1) identify natural resources within their boundaries; (2) provide for the preservation of land classified as mineral-resource areas until such time that regional needs dictate the conversion of such land to other uses; (3) determine on the master plan those lands that are available for immediate development or reclamation; (4) identify contiguous land which might be affected by surface mining; (5) establish time tables for the execution of local mining consistent with the projected needs of the community; and (6) provide for the protection of lands adjacent to mining operations through interior provisional zoning for the period of projected operations.

Id., page 23

After release of the Surface Mining Committee Report in 1970, the SMGB attempted to draft legislation to implement the committee's recommendations. Although several drafts were prepared, not one was presented to the legislature. It wasn't until a few years later, when Senator John Nejedly assumed leadership over the matter, that the committee's recommendations received the legislative consideration its authors had intended.

After another round of public hearings, visits to mine sites, and extensive analysis, many of the Surface Mining Committee's conclusions were finally incorporated into Senator Nejedly's SB 756, which as adopted became SMARA. Although the bulk of

The Surface Mining Committee's recommendations were incorporated into SMARA, adopted five years after the committee's report.

the bill's provisions addressed mine reclamation, discussed in detail in chapter 5, SB 756 also embarked California on a significant new path of attempting to identify and conserve its important mineral resources, enacting provisions described below.

Classification of Mineral Lands

Before local agency responsibilities to protect important mineral resources apply, it is only logical that resources must first be identified. This occurs through two processes: **classification** by the State Geologist, discussed below, and **designation** by the State Mining and Geology Board, discussed later in the chapter.

Overview of Classification

SMARA requires that OPR identify areas within the state that are urbanized or subject to urbanization or irreversible land uses that would preclude mineral extraction.

Fundamental to SMARA's mineral conservation provision is the Act's requirement that all mineral lands in California be classified. In implementing that ambitious goal, the Act provides that:

> On or after January 1, 1977, and, as a minimum, after the completion of each decennial census, the Office of Planning and Research shall identify portions of the... areas within the state which are urbanized or are subject to urban expansion or other irreversible land uses which would preclude mineral extraction[.]
>
> Pub. Res. Code § 2761(a)

Once OPR provides the information, the State Geologist must classify areas that contain mineral deposits based solely on geologic factors, not land ownership.

Once the Office of Planning and Research (OPR) provides that information, the State Geologist, based on SMGB guidelines, must then "on the basis solely of geologic factors, and without regard to existing land use and land ownership," identify: "(1) [a]reas containing little or no mineral deposits[;] (2) [a]reas containing significant mineral deposits[; and] (3) [a]reas containing mineral deposits, the significance of which requires further evaluation." *Id.* § 2761(b)(1)–(3).

CGS = California Geological Survey
OPR = Governor's Office of Planning and Research

Given the enormous collective size of California's mineral deposits, not all areas could be classified at once. Instead, the SMGB established priorities, first choosing sand and gravel resources in the fastest-urbanizing areas where the need was most immediate. These included the metropolitan regions of Los Angeles, San Francisco, and San Diego. Ralph C. Loyd, "Mineral Land Classification in California," *California Geology*, Mar. 1982, page 49. After initial priorities were identified, the SMGB prepared classifications for the lower Sierra Nevada range, the Sacramento area, the Central Valley, Inland Empire, and the North Coast. Initially, the areas were divided into **production-consumption regions,** but, since 1989 they have been classified by county.

Mineral classification information is reported in documents called "mineral land classification reports."

After holding a public workshop to allow discussion and comment on the geologist's findings, the California Geological Survey (CGS) (formerly known as the California Division of Mines and Geology) reports its findings relating to each region or county in a document called a **mineral land classification report.** This report forms the basis for potential designation of the studied lands as having significant mineral resources.

Criteria for Determining the Significance of Mineral Deposits

When preparing a mineral land classification report, CGS staff geologists conduct an investigation to determine the significance of the deposit being studied. This involves reviewing geologic and mining-related literature, compiling geologic maps, and plotting reported mines and prospects. Cal. Dep't of Conserv., *Guidelines for Classification and Designation of Mineral Lands,* page 3. The geologists also conduct field work including site investigations, rock sampling, surveying, and geologic mapping. *Id.*

When preparing mineral land classification reports, CGS staff conducts a geologic investigation involving a review of geologic and mining literature, compilation of geologic maps, and plotting of reported mines and prospects.

The geologists then apply economic criteria to determine if the area being studied meets SMARA's standard of significance. The first criteria is the deposit's marketability. To be significant, a deposit must be "minable [sic], processable, and marketable under the technologic and economic conditions that exist at present or which can be estimated to exist in the next 50 years." *Id.,* page 2. Moreover, "[b]ecause some of the conditions affecting extraction and marketability cannot be accurately projected 50 years into the future, conservative estimates will be made in assessing whether a particular mineral resource can be mined, processed, and marketed within the next 50 years." *Id.*

To be significant, a mineral deposit must be mineable, processable, and marketable under present economic and technologic conditions or the conditions estimated to exist in the next 50 years.

If it meets the marketability factor, a deposit's potential significance is then subject to the next criterion: whether it exceeds the threshold value that corresponds to the type of mineral or material being evaluated. For a deposit of construction materials—sand, gravel, and crushed rock—to be deemed significant, the deposit's total value must exceed $12,500,000 in 1998 dollars, adjusted annually for inflation (in 2005, that amount was $15,400,000). *Id.,* page 2. Other types of minerals have different threshold values. These values, in turn, "[are] based on the gross selling price of the first marketable product from an individual mineral deposit (or from a group of deposits that can be operated as a unit) after completion of extraction and any required mineral separation and processing." *Id.*

Mineral deposits must meet a threshold economic value—an aggregate deposit in 2005 had to exceed $15.4 million to be considered significant.

After the deposit under study has been evaluated, the geologist places the mineral deposit in a classification corresponding to its significance. The various classifications are discussed below.

Specific Mineral Land Classifications

MRZ-2a. An area meeting the economic criteria for significant mineral resources is classified MRZ-2a, the most deserving of conservation and protection. As the SMGB's Mineral Land Classification Guidelines note, "[a] typical MRZ-2a area would include an operating mine, or an area where extensive sampling indicates the presence of a significant mineral deposit [available for future exploitation]." *Id.,* page 4.

MRZ-2a is assigned to areas that have active mines or where extensive sampling indicates the presence of a significant deposit.

MRZ-2b. An area where the significance of a deposit is inferred, rather than actually measured, is classified MRZ-2b. This classification is also given to an area where mineral reserves have actually been demonstrated, but where those reserves cannot

MRZ-2b refers to areas where the significance is inferred or where reserves are demonstrated but do not meet the economic threshold of significance.

be determined to meet the economic criteria of significance. The Guidelines note that, for these areas, "[f]urther exploration work and/or changes in technology or economics could result in upgrading areas classified MRZ-2b to MRZ-2a." *Id.* The Guidelines go on to say that "[a] typical MRZ-2b area would include sites where there are good geologic reasons to believe that an extension of an operating mine exists or where there is an exposure of mineralization of economic importance." *Id.*

MRZ-3 resources have either a moderate potential for discovery of economic deposits or their existence is at least plausible.

MRZ-3a and MRZ-3b. While the previous two classifications involve clearly established mineral resources, MRZ-3, by contrast, is for an area that *could* potentially qualify as having mineral resources. An area classified MRZ-3a is believed to have a "moderate potential for the discovery of economic mineral deposits." *Id.,* page 6. For an area classified MRZ-3b, "geologic evidence leads to the conclusion that it is *plausible* that economic mineral deposits are present." *Id.* (emphasis added).

In MRZ-4 areas, knowledge regarding mineral occurrence is lacking.

MRZ-4. An area "where geologic information does not rule out either the presence or absence of mineral resources" is classified MRZ-4. *Id.* As the Guidelines note, "[i]t must be emphasized that [the] MRZ-4 classification does not imply that there is little likelihood for the presence of mineral resources, but rather there is a lack of knowledge regarding mineral occurrence." *Id.*

MRZ-1 areas either have no mineral resources or there is little likelihood they are present.

MRZ-1. Finally, an area is classified MRZ-1 "where adequate geologic information indicates that no significant mineral deposits are present, or where it is judged that little likelihood exists for their presence." *Id.,* page 4.

Special Classifications of Aggregate Resource Areas

An additional classification is an aggregate resource area that provides estimates of the area's 50-year supply of construction materials.

Areas that contain deposits of construction materials, or **aggregate,** may be classified MRZ-2a and MRZ-2b. Because of the compelling demand for these materials—used in roads, highways, and all types of construction—the areas are further separated into aggregate resource areas (ARAs). According to the Guidelines, these areas provide:

> [A] semi-quantified estimate of construction aggregate resources which are likely to be available to satisfy society's needs during the 50-year period following classification of an area. This estimate, when compared to [Division of Mines and Geology] projected needs for the next 50 years, provides the context for communities to plan for future aggregate needs in their land use policies.
>
> *Id.,* page 6

ARA = **Aggregate resource area**

To identify relative degrees of importance, deposits of construction materials are separated into one of three ARAs, as indicated below.

Immediately significant resource areas include existing aggregate mining operations.

Immediately significant area. All permitted mining operations within the area being studied are designated immediately significant. Guidelines, *supra,* page 6. Identification of all existing mining operations is important to ensure that agencies can plan for the uses surrounding those operations and that incompatible uses do not encroach on those operations.

Highly significant area. An area adjacent to an operating mine that has construction materials deposits or an area containing 10 times the threshold value of significance for aggregate is classified highly significant. *Id.* Its proximity to an existing operation makes extension of those operations and permitting of new mining operations more likely to occur. Accordingly, this classification should signal a lead agency that the area deserves immediate protection from encroachment from incompatible uses as well as consideration of and planning for use as a future mining site.

Highly significant resource areas are those that exceed 10 times the significance threshold for aggregate.

Significant area. An area with construction material deposits that does not meet any of the above criteria is classified simply as a significant ARA.

All other aggregate deposits are considered significant areas.

Excluded area. Certain deposits of construction materials are excluded from an ARA for economic or social reasons. Those excluded for economic reasons include:

Certain areas are excluded from ARAs for economic reasons, such as residential development, commercial areas, and public buildings and facilities.

- Residential areas and areas committed to residential developments
- Commercial areas with land improvements
- Industrial areas
- Major public or private engineering projects, including canals, freeways, bridges, airports, dams, railroads, major pipelines, and major power transmission lines

Id., page 7

Those excluded for social reasons include:

Additional areas are excluded for social reasons, such as cemeteries, parks, schools, and military bases.

- Cemeteries
- Public parks, developed historical sites and structures, and public recreation areas of all types
- Public or private schools, institutions, hospitals, and prisons, including adjacent development such as parking lots
- Military bases and reservations

Id.

Designation of Significant Mineral Resources

For planning purposes, the classification process is vital. It provides information about the quality and quantity of mineral resources that local agencies must have to protect and conserve those resources. As explained in the next section, classification triggers certain responsibilities with which a lead agency must comply when making a land use decision. Yet, it is also possible to go one step beyond classification. Under SMARA, the SMGB may further designate an area classified as MRZ-2a or MRZ-2b as an area of statewide or regional significance. Pub. Res. Code § 2790; Cal. Dep't of Conserv., *Guidelines for Classification and Designation of Mineral Lands*, page 8. When this happens, the agency must accord the area additional protection than the classification alone provides.

After an area is designated MRZ-2a or MRZ-2b, the SMGB may designate it as either a significant regional or statewide resource.

SMARA defines an **area of regional significance** as:

[One]...known to contain a deposit of minerals, the extraction of which is judged to be of prime importance in meeting future needs for minerals in a particular region of

> the state within which the minerals are located and which, if prematurely developed for alternate incompatible land uses, could result in the permanent loss of minerals that are of more than local significance.
>
> Pub. Res. Code § 2726

The Act defines an **area of statewide significance** as:

> [One]...known to contain a deposit of minerals, the extraction of which is judged to be of prime importance in meeting future needs for minerals in the state and which, if prematurely developed for alternate incompatible land uses, could result in the permanent loss of minerals that are of more than local or regional significance.
>
> *Id.* § 2727

Before designating an area a significant mineral resource, the SMGB must seek the recommendations of affected agencies and hold a public hearing.

Prior to designating an area one of regional or statewide significance, the SMGB must hold a public hearing. *Id.* § 2790. It must also seek the recommendations of concerned federal, state, and local agencies as well as interested organizations and educational institutions. *Id.* § 2791. After the hearing, if the SMGB decides to make that designation, it must do so by regulation and indicate:

> [T]he reason for which the particular area designated is of significance to the state or region, the adverse effects that might result from premature development of incompatible land uses, the advantages that might be achieved from extraction of the minerals of the area, and the specific goals and policies to protect against the premature incompatible development of the area.
>
> *Id.* § 2790

These policies must be explained in a formal designation report that must also must identify the boundaries of the designated area, the specific deposit being designated, and an estimate of the commodity quantity available for mining under present or foreseeable technologic, economic, and land use conditions. *Guidelines*, *supra*, page 8. The designation report is critical as it provides the information a lead agency needs to plan for and protect the designated resources.

City and County Responsibilities for Protecting Classified and Designated Lands

SMARA imposes policy- and project-level obligations on cities and counties when dealing with significant mineral resources.

The classification and designation processes described would be mostly meaningless if no consequences followed. While knowledge of the location and extent of mineral resources might be valuable in its own right, that knowledge by itself would not protect and conserve the resources. SMARA, therefore, requires that cities and counties not only recognize that certain lands within their jurisdiction contain important mineral resources, but also utilize that information to carry out policies of the Act. It does this by (1) mandating policy-level planning with the requirement that agencies incorporate policies to protect mineral resources in the general plan; and (2) requiring that an agency give special consideration to mineral resources when reviewing a project involving those resources or one that threatens their potential extraction.

Policy-Level Requirements

To achieve its mineral conservation policies, SMARA first requires that cities and counties adopt mineral resource management policies whenever they receive a mineral classification report or whenever the SMGB designates land within their jurisdictions as an area of regional or statewide significance. Pub. Res. Code § 2762(a). Local agencies must include these policies within their general plans and must submit those policies to the SMGB for review. *Id.* § 2762(b), (d); *see also* Gov't Code § 65302(d) (requiring that cities and counties adopt conservation elements in their general plans for "conservation, development, and utilization of natural resources including... minerals...."). The policies must: (1) recognize mineral classification information; (2) assist in the management of land uses affecting areas of regional and statewide significance; and (3) emphasize conservation and development of mineral deposits the classification process has identified. *Id.* § 2762(a)(1)–(3).

Cities and counties must adopt mineral resource management policies in their general plans that recognize mineral classification information, assist in management of land uses affecting mineral resources, and emphasize mineral conservation.

The SMARA regulations elaborate on these requirements, stating that city and county general plans must either (1) provide a summary of the information contained in the mineral land classification and/or designation reports; or (2) incorporate those reports by reference along with SMARA's legislative intent and state policy provisions. Cal. Code Regs. tit. 14, § 3676(a). General plans must also refer to the location of identified mineral deposits and must discuss those areas where possible future extraction may occur. *Id.* § 3676(c)(1). That same information should likewise be depicted graphically on overlay or planning maps. *Id.* § 3676(c)(2).

City and county general plans must summarize mineral land classification information, refer to SMARA's legislative intent provisions, and refer to and depict the locations of future extraction areas.

In addition to the information required in a general plan, the regulations require that agencies further implement at least one of the following measures:

- Use overlay zones, mineral resource /open space zones, or similar zoning designations identifying the presence of identified mineral deposits and restricting encroachment of incompatible land uses in the conserved areas. *Id.* § 3676(c)(3)(A). For purposes of these regulations, incompatible land uses are those that "may prevent mining because of the greater economic value of the land and its improvements. Examples of such uses may include, but shall not be limited to, high density residential, low density residential with high unit value, public facilities, geographically limited but impact intensive industrial, and commercial." *Id.* § 3675. In contrast, examples of uses the regulations deemed compatible with mining are "very low density residential, geographically extensive but low impact industrial, recreational, agricultural, silvicultural, grazing, and open space." *Id.*
- Record notices of the presence of identified mineral depositions on property title in affected mineral areas. *Id.* § 3676(c)(3)(B)
- Impose conditions in the general plan upon incompatible land uses and the mineral deposit in surrounding areas to require mitigation of land use conflicts prior to approving a use incompatible with mining. *Id.* § 3676(c)(3)(C); *see also* § 3675 (defining uses "compatible" and "incompatible" with mining).

Cities and counties must implement at least one of the following measures: (1) placing classified or designated areas into special or overlay zoning districts, (2) recording notices of the mineral resource in affected areas, or (3) creating special general plan policies to mitigate land use incompatibilities between mining and non-mining uses.

Project-Level Requirements

Beyond requiring the policy-level planning described above, SMARA imposes project-level requirements for certain land use applications affecting classified or designated mineral resources. These requirements apply to any application that "threatens the potential" to extract from a mineral resource or, for designated lands, to any project that involves the mineral resources themselves.

SMARA imposes project-level requirements when cities and counties consider projects that threaten the potential to extract minerals or projects involving the resources.

Application of mineral management policies to land use projects that "threaten the potential" to extract from classified or designated mineral lands. Cities and counties have special responsibilities whenever they review individual projects that "threaten the potential to extract minerals" from lands that have been classified or designated as important mineral resources. Pub. Res. Code §§ 2762(a), (b), 2763(a). These applications typically involve projects adjacent to those lands, such as projects that would allow neighboring residences, businesses, or other non-mining uses.

A precise definition of the term, "threaten the potential to extract minerals," is elusive, but certainly includes incompatible uses proposed on the mineral lands; what other lands are included is not clear.

Identifying projects that threaten the potential to extract minerals. Unfortunately, because SMARA does not define "threaten the potential to extract minerals," the types of projects within the ambit of the term cannot precisely be identified. First, where must a project be located to have the potential to prevent extraction? The phrase might include any proposed non-mining project on land containing the deposits. Certainly, such a project would directly remove mineral lands from production due to incompatible development. But would other lands also fall within the term?

As mentioned above, the history surrounding SMARA's adoption indicates that the moving force behind the legislation was a broad concern about the effects of urban encroachment on present and future mining sites. Consistent with that understanding, a logical interpretation of the term suggests that agencies must consider the importance of mineral deposits whenever residential, commercial, or other sensitive uses are proposed adjacent to lands with mineral deposits. A broader interpretation suggests that agencies must even consider the threatening potential of a project miles away so long as the project might arguably affect future mineral extraction.

The construction of a residential subdivision in the vicinity of a mineral deposit, for example, might threaten future extraction since the impacts—from dust, noise, or blasting waves—would be experienced far beyond properties next to the mine. Similarly, if only certain roads are available to transport materials from a future site, the approval of development along those roads would likewise threaten future extraction. Indeed, a common complaint about mining operations relates to the large number of heavy trucks needed to transport the materials. Thus, under the broadest interpretation of the term "threatens the potential to extract minerals," any development along such roads might trigger a lead agency's duty to consider the effect on mineral projects when approving that development.

Unfortunately, without a definition, a precise interpretation of the term's geographic reach is elusive. Absent that definition, cities and counties should interpret "threatens the potential to extract minerals" to include *at least* those projects proposed on lands adjacent to classified or designated lands. And to be safe, agencies should also include those projects on non-adjacent lands for which there is a reasonable basis—supported by credible evidence, not simply speculation—that development might hinder extraction.

This raises an additional issue: what *specific types* of projects potentially hinder extraction? Again, because the term is not defined, SMARA does not answer the question. Nonetheless, SMARA's regulations do define—in a related context—ones that are "compatible" and "incompatible" with mining: "Compatible" uses include "very low density residential, geographically extensive but low impact industrial, recreational, agricultural, silvicultural, grazing, and open space." Cal. Code Regs. tit. 14, § 3675. "Incompatible uses," by contrast, include "high density residential, low density residential with high place value, public facilities, geographically limited but impact intensive industrial, and commercial." *Id.* Thus, in the absence of a precise definition, agencies should interpret the term to encompass those uses the regulation deems "incompatible" with mining.

Requirements for projects that would threaten the potential to extract from mineral lands. When considering a project that "threatens the potential to extract minerals" on land that has been classified or designated, a city or county must prepare and transmit a statement to the State Geologist and SMGB indicating the reasons for permitting that use. Pub. Res. Code § 2762(d). SMARA, however, does not prohibit a city or county from permitting the proposed use. Nor, by requiring its transmittal, does SMARA grant either the power to veto or modify an agency's decision. SMARA's intent, rather, is to require that the agency at least recognize that the project being considered may affect the long-term ability to exploit important mineral deposits.

In carrying out this duty, cities and counties must include their statements of reasons "in conjunction with any environmental document required by [the California Environmental Quality Act (CEQA)]." *Id.* If CEQA applies, the agencies must further "comply with the public review requirements of that [Act]." *Id.* Consequently, cities and counties must circulate their statements for permitting the use in any categorical exemption, negative declaration, or environmental impact report prepared for the project under review. Only by doing so can cities and counties guarantee that the statements will be circulated for public review and comment and that the agency will respond to all comments, as required under CEQA.

Application of mineral resource management policies to land use projects "involving" lands designated as significant mineral resources. The responsibility of a city or county to conserve resources does not apply only to projects that threaten the potential to extract from mineral resources. For lands designated (and not simply classified) as significant mineral resources, the responsibility also applies to projects "involving" the resources themselves—that is, to mining projects proposed on mineral lands. While the former type only requires that an agency notify the appropriate

PRACTICE TIP

In the absence of a precise definition, "threaten the potential to extract minerals" should be interpreted to include at least those projects proposed adjacent to affected mineral lands that meet the regulatory definition of "incompatible" use. Other lands can also fall within the term if there is a reasonable basis for concluding that nearby development would hinder the potential extraction of the mineral lands.

CEQA = California Environmental Quality Act

PRACTICE TIP

A city or county should determine through an initial study whether a project that threatens the potential to extract from mineral resources might create significant and adverse environmental impacts. If there is a potential for such an impact, the project requires preparation of an EIR, or a mitigated negative declaration, to ensure that the agency will properly consider and mitigate any impacts to the mineral resources.

It is also important to inform the State Geologist and SMGB of the project as soon as possible. Although SMARA does not make clear when notification is required, an agency should contact them at an early stage, preferably by forwarding copies of the project's initial study.

state agencies and provide a statement of reasons for approving the project in accordance with CEQA, the latter has a greater impact on a city's or county's land use authority. When considering a project that involves designated mineral lands, the agency's decisions must also be "in accordance with... mineral resource management policies." *Id.* § 2763(a), (b). In "balancing mineral values against alternative land uses," the decisions must take into account the project's importance not only to the cities and counties affected, but to the market region for the minerals (for deposits of regional significance) or to the state and nation (for deposits of statewide significance). *Id.*

In this respect, SMARA affects the authority of a city or county to regulate land use. As noted in chapter 3, SMARA expressly states that it does not limit an agency's zoning and police powers. Pub. Res. Code § 2715(a), (f). SMARA therefore does not—and by its very terms, cannot—mandate the decision a city or county must make when considering projects involving designated lands. Through the mineral conservation provisions, however, SMARA certainly encourages cities and counties to approve extractive uses on designated lands. When they choose not to do so, evidence must exist that their decisions are nonetheless in accordance with their mineral resource management policies and that they engaged in the requisite balancing of mineral values versus alternative land uses. If they fail to comply with these obligatins, cities and counties could violate their general plans and be subject to appeals to the SMGB, as described below.

A city or county's denial of projects involving designated resources must be consistent with its mineral resource management policies and reflect the requisite balancing of mineral values versus alternative land uses.

Designation Appeals to the State Mining and Geology Board

Some decisions involving applications to state-designated mineral resources can be appealed to the SMGB.

After land has been designated as a significant statewide or regional resource, the SMGB may hear an appeal of certain land use decisions the city or county makes regarding that land. A **designation appeal** allows the SMGB to review the city or county decision that approved or denied a permit to mine the land's significant mineral resources. Pub. Res. Code § 2775(a). In hearing the appeal, however, the SMGB does not have the full appellate power of a court to review other land use decisions. A designation appeal only allows the SMGB to determine "whether the [city's or county's] decision is supported by substantial evidence in light of the whole record." Pub. Res. Code § 2775(c). And if it finds that such evidence does not exist, the SMGB can only direct the city or county to "schedule a hearing to reconsider its action;" it cannot overturn the decision. *Id.*

The designation appeal is a seldom used, little understood process, and the provisions governing the appeal raise a number of unresolved issues. For example, what specifically should the SMGB review? SMARA empowers the SMGB to determine if substantial evidence supports an appealed decision but, curiously, does not specify the aspect of the decision that requires substantial evidence. This is significant since approval or denial of a mining project usually entails several findings, many of which a city or county must make not based on any requirement of SMARA, but on the unique requirements of a general plan, zoning ordinance, and other local land use requirements. Does the designation appeal give the SMGB authority to review those findings as well?

Although not spelled out, the answer is, probably not. Other provisions of SMARA confirm that the Act does not limit a city's or county's police and zoning powers. Pub. Res. Code § 2715(a), (f). The matters that the legislature intended to be subject to a designation appeal, therefore, are likely only those related to a city's or county's responsibilities to conserve mineral resources.

SMARA's designation appeal provisions probably confine the SMGB to reviewing only those matters relating to the city or county mineral conservation responsibilities.

An additional unresolved question concerning the designation appeal is, what types of city or county decisions may the SMGB review? The language of SMARA seems plain, stating that the SMGB may review an approval or denial of "a *permit* to conduct surface mining operations." *Id.* § 2775(a) (emphasis added). SMARA would thus seem to allow the SMGB to review only a mining application that involves a conditional use permit. Yet, a mining project may involve more than an application for a permit; often together with or as a precursor to a permit application, other entitlements are needed, such as an amendment to the general plan or a zoning change. *See* chapter 7 (discussing the general plan and zoning ordinances).

While the language in SMARA appears to exclude such an application from the ambit of a designation appeal, the exclusion seems odd given the responsibilities SMARA assigns to cities and counties to conserve designated mineral resources. In some cases, mining cannot take place on designated land unless an agency approves a general plan amendment or zoning change. So, it's reasonable to assume that the SMGB—in its role as monitor of city and county responsibility to conserve designated resources—would have equal power to review a decision relating to that entitlement. But, because the language of SMARA confines the SMGB's review only to permit decisions, the legislature apparently intended otherwise.

Although mining applications may involve a number of entitlement requests, the language of SMARA allows the SMGB to hear appeals only of decisions granting or denying permits.

A third issue relates to the remedy the SMGB may order as part of a designation appeal. SMARA provides that the SMGB, when granting an appeal, "shall remand the appeal to the [city or county] and the [city or county] shall schedule a public hearing to reconsider its action." *Id.* § 2775(c). That the SMGB may only order reconsideration raises a question about what significance granting an appeal has on the appealed decision. If the SMGB orders reconsideration, how must the city or county respond? Must it reweigh the evidence? Need it change its decision?

If the SMGB grants the designation appeal, it cannot overturn the appealed city or county decision; it can only remand the matter back to the city or county for reconsideration.

The answers to the last two questions are, probably not. The SMGB's power in a designation appeal is intentionally limited. As explained in chapter 5, the SMGB, through a reclamation plan and financial assurance appeal, can overturn a city or county reclamation plan and financial assurance decision. That the legislature chose not to give the SMGB that same power in the designation appeal indicates an intent to limit the SMGB's ability to interfere with a city or county decision in a designation appeal.

Because the SMGB is the chief policy-making body for SMARA, its resolution of the designation appeal would likely be persuasive to a court considering a lawsuit that raises the same or similar issues.

Nonetheless, an order to reconsider would likely have an important practical effect. When challenging a city or county permit decision, an aggrieved party would probably not only file an appeal to the SMGB; it would probably also file a lawsuit in which many of the same issues would be raised. The SMGB appeal might be resolved even before the lawsuit could be heard on its merits, since judicial dockets in almost every California court jurisdiction are overcrowded. And, because most judges have little prior experience either with SMARA or its mineral resource conservation provisions, the SMGB's resolution of the issues presented in the designation appeal are likely to be very persuasive. Indeed, it seems natural that a judge might defer to a state agency with the most knowledge of the issues and that also has policy-making authority over the Act. Practically speaking, therefore, the SMGB appeal can significantly influence the result of any lawsuit premised on the same grounds, even if the SMGB's power in such an appeal is limited.

Parties filing a designation appeal must do so within 15 days of exhausting their rights with their cities or counties; normal administrative appeal procedures then apply.

To perfect such an appeal, an aggrieved party must file a notice of appeal within 15 days of exhausting the right to appeal under city or county policies and procedures. *Id.*; Cal. Code Regs. tit. 14, § 3623(a). The applicant must also submit a map to the SMGB indicating the exact location of the proposed mining operation, written statements with supporting documentation of the basis for the appeal, and a copy of a notice of intent to appeal that the applicant has filed with the city or county. Cal. Code Regs. tit. 14, § 3623(a)(1)–(3). Once all the information is filed, normal administrative procedures govern the appeal. An administrative record is prepared and a public hearing is held that follows normal administrative procedures. Cal. Code Regs. tit. 14, §§ 3632–3633. The SMGB's practice is to hold the hearing somewhere within the territory of the city or county whose decision is under appeal.

in this chapter...

chapter 5

Reclamation

Surface mining is unique among land uses. Not only is the way in which it occurs regulated, but a special set of regulations governs the way mining must conclude. This regulation of mine closure, known as reclamation, is necessary because surface mining creates significant land disturbance, often leaving the property with deep pits and other land forms unsuitable for subsequent land uses. Unless the land is to be left in a mined-out condition—with all the environmental, aesthetic, and public safety problems that would create—something must be done to restore the property when mining is complete. Stated more simply, to prevent mining from leaving a permanent scar on the landscape, mined land must be reclaimed.

In California, the State Surface Mining and Reclamation Act of 1975, or SMARA, governs reclamation of mines. This chapter sets forth SMARA's sometimes complex, technical requirements, as well as the reasons for reclamation and its benefits. In addition, this chapter also presents the history leading to the law's adoption and the significant changes to SMARA that followed.

SMARA = Surface Mining and Reclamation Act

Why Reclamation?

Unlike other industrial uses, which operate *on* the land, mining *consumes* the land. Indeed, the quality and quantity of its deposits is the very reason mine property is selected. In exploiting mineral depositions over a long period—sometimes spanning decades or more—a mine operator will usually disturb a huge amount of land, both vertically and laterally. Consequently, what may originally have been scenic or pastoral land may, after many years of mining, resembles a moonscape with pits many hundreds of feet deep and several acres wide. Allowing a mine operator to leave property in this condition makes no sense. Mined-out property is generally

Over the lifetime of a mining operation, the operator may disturb huge amounts of land, both vertically and horizontally.

If left in its mined-out condition, the property would remain in an aesthetically unpleasing, unsafe, and environmentally-damaged condition.

not aesthetically pleasing, especially from the standpoint of nearby residents who would have to look upon it permanently if reclamation did not occur. Steep slopes and deep pits resulting from years of mining may also make the land unsafe and create attractive nuisances. Mining may further disturb sensitive habitat, altering a natural water course or drainage pattern, or cause contamination from chemical products used during operations.

Given these aesthetic, safety, and environmental concerns, mine operators simply cannot be permitted to walk away once mining is done. Yet, that has happened on numerous occasions in the past. According to an Office of Mine Reclamation report, California has approximately 39,000 abandoned mines, 84 percent of which present physical hazards and 11 percent environmental hazards. Cal. Office of Mine Reclamation, *California's Abandoned Mines*, page 8 (June 2000). While many were small, historic operations for mining gold, a few were very large commercial operations.

To prevent the numerous public problems abandoned mines create, California requires that all mining operations be reclaimed. Reclamation ensures that taxpayers do not bear the costs of mine site remediation.

To prevent abandonment and the numerous public problems it creates, SMARA now requires that every operator reclaim the mined-out land for productive post-mining use. As part of the process, an operator must remediate any remaining adverse conditions and prepare the property for reuse. Reclamation guarantees the elimination of all post-mining externalities. Rather than force aesthetic, public safety, and environmental costs of unreclaimed mine property on the public, a mining operator is required to bear the cost. *People ex rel. Dep't of Conserv. v. El Dorado County* (2005) 36 Cal. 4th 971, 990–91.

Because mining takes place in a variety of geographic, geologic, and environmental settings, the manner of reclaiming an operation will vary. But, generally speaking, reclamation requires that an operator backfill or reshape pits and excavations with suitable engineered material, taking care to recontour, revegetate, and prepare the property to support a proposed post-mining use. Reclamation also requires the removal of plants, equipment, and facilities used during mining, the restoration of disturbed wetlands and habitat, and remediation of any adverse environmental conditions that remain. *See* Pub. Res. Code § 2733 (defining **reclamation**). The thrust of reclamation is to prepare mined property for a subsequent land use that, as discussed below, might include a park, open space, recreation, housing, or continuing industrial use.

Reclamation requires that mined property be prepared for a post-mining land use, which involves land reshaping, revegetation, restoration, and recondition.

State law imposes two basic requirements to ensure reclamation: operators must have a reclamation plan and must post financial assurances.

To accomplish reclamation, state law imposes two basic requirements. First, an operator preparing a property for post-mining use must have a comprehensive reclamation plan identifying measures that will be taken to restore the property and prepare for its subsequent use. Second, the operator must post some form of financial assurance, such as a bond or letter of credit, to guarantee that the property will be reclaimed in accordance with the plan. In addition, operators have other responsibilities, including the requirement that they submit annual mining reports and be subject to annual inspection. These and other responsibilities are discussed below and in chapter 6.

Another View of Reclamation—The Second Use Concept

While necessary for restoring mined-out property, reclamation offers more than just a means to address post-mining externalities. Under SMARA, every operator must identify in the reclamation plan an **end use** for the property after mining. This **second use** requirement presents a unique opportunity to convert the mined-out property to attractive, lucrative land uses involving any number of possibilities—commercial, industrial, residential, recreational, open space, natural preserve, and many others.

Reclamation should be viewed as more than just a means for dealing with post-mining externalities. Reclamation also allows mine property to be put to a second use that can involve a number of land uses.

Several surface mines have already been reclaimed for a valuable second use. Larkspur Landing in Marin County is a prime example. Once a rock quarry, the property has been transformed to an attractive, mixed-use development with a major regional ferry terminal, office park, shopping center, and housing site. Elsewhere in California, mining operations have been converted to farmland, residential developments, golf courses, and habitat preserves.

Outside California, a former quarry bowl in Sparks, Nevada has been filled with water to become a marina. Visitors to Victoria, British Columbia may also be familiar with the famous Butchart Garden, a quarry converted to beautiful gardens that attract visitors from around the world.

As these examples illustrate, reclamation can be very profitable for an imaginative operator who wishes to maximize the value of mined-out property. This is especially true since reclamation may not occur until many years after plan approval. In the period preceding reclamation, a mine operator can forge relationships with land developers, obtain necessary capital, and build consensus among local leaders, neighborhood groups, and interested organizations for a proposed second use. In this way, an operator can sometimes lock in the intended use long before the conclusion of mining.

Reclamation can be very profitable for imaginative mine owners, allowing them to effectively lock in their post-mining land uses well before mining concludes.

Reclamation can be beneficial to cities and counties as well when the preceding period is used to ensure a second use responsive to community needs. A former mine site may provide an excellent opportunity for infill or make possible new commercial development and office space. A reclaimed site can also be converted to a natural use such as open space, wildlife preserve, and recreation.

Reclamation offers cities and counties the opportunity to convert mine property to a well-planned land use that responds to community needs.

Given the potential secondary uses, therefore, reclamation should not only be seen as a vehicle for dealing with post-mining externalities. With its possibility of a second life, mined property should be thought of as having dual value, providing the potential for a beneficial and well-planned land use. Mine operators and local agency officials who recognize this can benefit greatly, each in their own particular way.

History of Reclamation in California

Mining has been an important industry in California since before the state was founded. California's existence owes much to the discovery of gold in the late

Top left: From 1924 until the 1970s, the Hutchison Company operated a large rock quarry in Marin County. Rock from this operation was used in fortifying Treasure Island, in creating access roads for the Golden Gate Bridge, and for many other projects throughout the Bay Area.

Top right: A proposed zoning map prepared prior to the reclamation of the Hutchison Quarry shows the post-mining land uses proposed for the property following reclamation.

Middle left: A 1991 photograph shows the Hutchison Quarry property following reclamation and annexation by the City of Larkspur. The former quarry property is now the site of the Larkspur Landing Shopping Center, a Golden Gate Ferry terminal, offices, hotels, and residential development. Portions of the old quarry bowl can still be seen in the background.

Bottom left: A conceptual drawing shows how commercial bayfront development might look following reclamation of the Hutchison Quarry.

1840s and the Gold Rush that followed. Yet, despite mining's longevity in the state, mandatory reclamation is a relatively recent development that began with the adoption of SMARA in 1975. And, even after that, local enforcement was spotty. Only in the past decade and a half—after significant revisions to the law and the adoption of a host of administrative regulations—has reclamation begun to be enforced with consistency and regularity.

SMARA's Antecedents

Reclamation as a private, rather than publicly mandated measure began as early as the nineteenth century when English landowners required mine operator lessees to replace topsoil and prepare mined land for agriculture after completion of mining. In the United States, reclamation began in the 1920s as a experimental program to convert West Virginia coal mines into quick-growth forests. After early successes with these programs, other operators expanded the scope of reclamation by converting their lands to pastures and row crop farm land. Zoe Mankoski and Kenneth Cole, "Reclamation Planning in California," *California Geology,* page 55 (Mar. 1982).

Reclamation, which began in England as a private measure, came to the U.S. in the 1920s, initially as an experimental program to convert coal mines to forest land.

As a result of these private initiatives, the first mandatory reclamation laws were adopted in the late 1930s. Confined mostly to eastern states, the laws generally pertained to coal mining and were concerned with remediating adverse environmental conditions remaining on the affected mine properties. *Id.* In response to growing public concern for the environment, a second wave of states began adopting reclamation laws in the mid-1960s and early 1970s. Report of the Committee on Surface Mining, page 8 (Oct. 2, 1970). By the time of SMARA's passage in 1975, California was one of the last states to adopt a mandatory reclamation law. Assembly Committee on Resources and Land Use, Analysis of SB 756, page 1 (Aug. 5, 1975).

The first mandatory reclamation laws were adopted in the late 1930s. California was one of the last states to adopt such a law, which it did in 1975 through SMARA.

Despite the absence of a mandatory state law, a number of California cities and counties had adopted reclamation laws prior to SMARA, with Los Angeles County being the first. James Pompy, "Mined Land Reclamation Program," *California Geology,* Jan. 1987, page 4. Many local agencies were reluctant to pass their own laws, however, because they lacked expertise in reclamation. As a result, a large majority of property in California went unreclaimed after being mined. Analysis of SB 756, *supra*, page 1.

Few local agencies had reclamation requirements before SMARA. As a result, prior to the Act's adoption, many mine properties went unreclaimed.

Adoption of SMARA

Recognizing that the mounting acreage of unreclaimed mines presented a serious problem, but lacking the expertise to address the problem, local leaders pressured lawmakers to adopt a mandatory state program. James Pompy, "Mined Land Reclamation Program," *California Geology,* Jan. 1987, page 4. In 1966, the legislature responded by passing a resolution calling upon the Senate Natural Resources Committee to consider "the subject of uniform controls and standards for strip mining." Report of the Committee on Surface Mining, page 1 (Oct. 2, 1970). The legislature, in turn, asked the State Mining and Geology Board (SMGB) to advise in that review.

In 1966, the legislature adopted a resolution to explore mining-related issues and asked the SMGB to advise.

SMGB = State Mining and Geology Board

Following its legislative directive, the SMGB formed an ad hoc Surface Mining Committee in 1968 that consisted of mining industry representatives and state and local officials. *Id.*, page 3. The committee conducted two years of public hearings, mine site visits, and other investigations. The scope of its work was broader than the subject of reclamation and also addressed mineral conservation issues, discussed

In accordance with the legislature's recommendation, a special committee suggested a fundamental state policy directed toward reclamation of mined land.

in chapter 3. After completing its analysis, the committee issued a report in 1970 recommending that "there be devised a fundamental State policy directed toward [among other things] *the reclamation of mined land*." *Id.*, page 22 (emphasis added).

Following receipt of the report, the legislature requested that the SMGB develop surface mining legislation to implement the recommendations. An ad hoc committee of the SMGB then prepared nine separate drafts of the legislation, although none reached the legislature in the years that followed. Letter to Governor Brown from Senator John Nejedly 1 (Sept. 15, 1975). It was not until Senator Nejedly asserted leadership over the matter in 1974 that significant progress was made. After conducting additional hearings that same year and in 1975, the Senator introduced SB 756 during the 1975 legislative session. That bill, with several new amendments, ultimately was adopted as SMARA, which became effective January 1, 1976.

SMARA was adopted in 1975, following the leadership of Senator John Nejedly.

Early Experience with SMARA's Reclamation Provisions

As initially adopted, SMARA's reclamation provisions were modest. They had only one basic requirement—that operators obtain approval of a reclamation plan.

As initially enacted, SMARA's reclamation provisions were modest. In mandating reclamation, the Act had only one basic provision: every mine owner had to receive local lead agency approval for a reclamation plan. Although also specifying the plan's minimum contents, the Act offered little additional detail or guidance. Absent, for instance, were any provisions specifying the consequences for failure to receive approval of a reclamation plan or failure to comply. Likewise absent were provisions mandating how a lead agency was to oversee and enforce operator compliance. Additionally, no financial assurance mechanism existed at the time to guarantee that an operator would follow through on its obligation.

SMARA's initial modesty was by design. As adopted, the Act was premised on a policy of home rule.

Despite the discovery of these shortcomings after the fact, SMARA's initial modesty was by design. The state's expertise was necessary to formulate reclamation policy, but those responsible for passage of the Act wanted to ensure that local agencies remained responsible for enforcement. As Senator Nejedly confirmed in a letter to the League of California Cities shortly before SMARA's adoption, the Act "does not contain enforcement provisions... because the bill provides for a *local* regulatory program. Enforcement provisions would be embodied in local ordinances." Letter from Senator John Nejedly to David F. Beatty, League of California Cities 1 (August 26, 1975) (emphasis in original). As initially adopted, therefore, SMARA was premised on a policy of home rule.

The flaws of that policy soon became apparent. By 1989, the California Department of Conservation found that only 11 percent of the mines covered since SMARA's adoption had actually been reclaimed in accordance with their reclamation plans. Assembly Committee on Natural Resources, Analysis of AB 3551, page 4 (Apr. 2, 1990). The study further found that 30 percent were either totally unreclaimed or only partially reclaimed. *Id.* The study noted not one documented instance in which a lead agency had penalized an operator for noncompliance. *Id.*

Moreover, at the time of the study few agencies had been regularly inspecting mining operations for compliance. Many "vested" or nonconforming operations had also failed even to submit a reclamation plan for lead agency approval.

In short, while retaining control over reclamation, local agencies generally were not successful in exercising their authority after SMARA's enactment due to lack of oversight or resources, or for other reasons. This early experience paved the way for significant revisions to the policy of home rule upon which the Act was founded.

Local agencies did not adequately enforce SMARA's reclamation requirements in the early years following its adoption. Their deficient enforcement paved the way for significant revisions to the Act.

1990s Amendments to SMARA

Recognizing the problems created by SMARA's modest provisions, the legislature enacted revisions to the Act in the early 1990s. Authored by Senator Byron Sher, these revisions created provisions for administrative penalties exposing a noncompliant operator to potentially significant consequences. The revisions further required that an operator submit an annual compliance report and post a financial assurance to guarantee that the mine site would be reclaimed. An operator who did not prepare a reclamation plan or post a financial assurance could also be prohibited from selling material to any state agency.

SMARA was significantly reformed in the early 1990s to require that operators post financial assurances, submit annual reports, and comply with additional duties.

The revisions also imposed significant enforcement responsibilities on cities and counties. A lead agency had to conduct an annual inspection of every mining operation within its jurisdiction, and new administrative enforcement and civil penalty mechanisms to guide an agency through the process were added. Any agency that failed to comply with the provisions could be relieved of some or all of its responsibilities by the SMGB under the Act's "takeover provisions" explained in chapter 6.

SMARA's reforms imposed new responsibilities on cities and counties, including the duty to annually inspect mining operations.

A Greater State Role

As a result of the 1990s revisions, SMARA gained the "teeth" that it did not initially have. Since these revisions, the SMGB has acted proactively in recent years to enforce operator and lead agency compliance and, in some cases, has even assumed enforcement authority over cities and counties whose oversight was thought to be deficient. In addition, the SMGB has promulgated new regulations to enforce the Act, some generating opposition and lawsuits from the mining industry.

Following revisions to SMARA, the SMGB has taken a more pro-active role in adopting administrative regulations and enforcing the Act.

The tension between SMARA's transition from a purely home rule statute to one requiring that enforcement power be shared by local government and the state came to the forefront in the 2005 legislative session. At that time, the legislature was considering SB 668, a bill opposed by the mining industry, that would have granted significant authority to the California Department of Conservation. The bill would have allowed the Department to seek judicial relief whenever a lead agency approved a deficient reclamation plan. The bill would have also permitted the Department to seek an injunction against an operator who deviated from its reclamation plan. The proposed language in SB 668 was occasioned by a contentious dispute about jurisdicton that had wound its way to the California

Supreme Court. In light of the Court's decision in that case, *People ex rel. Department of Conservation v. County of El Dorado* (2005) 36 Cal. 4th 971, which ruled that the Department already had such powers, the author must have dropped SB 668 as no longer necessary.

While SMARA initially vested cities and counties with exclusive enforcement authority, the Act and its regulations now provide for an expanded state enforcement role.

Nonetheless, this episode illustrates how much the Act has evolved since initial adoption, both as written and as interpreted. While SMARA's provisions were at first modest, vesting local agencies with exclusive authority over reclamation enforcement, the Act and its administrative regulations now provide for an expanded and important state enforcement role.

This was recently confirmed by a court decision that arose from another contentious dispute between the mining industry and SMGB. After the SMGB adopted a regulation requiring its concurrence that an operator had completed reclamation before release of the operator's financial assurance, a mining industry association sued, claiming the regulation exceeded the SMGB's authority. The Sacramento-based Third District Court of Appeal disagreed, building on the Supreme Court's *El Dorado* decision. As the Court observed:

> The State Supreme Court's pronouncements in *El Dorado* refute the Associations' claim that the Legislature relegated the Director [of the Department of Conservation] to a *mere* advisory role in achieving SMARA compliance. Although, as a general principle, the Director has a secondary role when compared to [cities' and counties' roles], there is no doubt that the Director has important statutorily rooted responsibilities to ensure that reclamation is completed satisfactorily and that financial assurances are adequate to cover the cost.
>
> *Minerals Ass'ns Coalition v. State Mining & Geol. Bd.* (2006) 138 Cal. App. 4th 574, 585

The *Mineral Associations* case amply highlights how SMARA has emerged from its home rule origin. Although SMARA initially relegated the state to purely a policy-making function, the Act's 1990s amendments created a dual local-state enforcement scheme that, while still giving a local agency primary authority, nonetheless provides for an important state role.

SMARA Function and Applicability

Having explained the reasons for reclamation, its benefits, and the history of implementation through SMARA, this chapter now reviews the substantive requirements of reclamation. To begin at the most logical starting point, the entities, land, and activities that SMARA subjects to reclamation are discussed first.

Activities, Lands, and Entities Subject to Reclamation

Surface mining operations. SMARA's reclamation provisions apply to any person or entity engaged in a **surface mining operation**. This includes:

> [A]ll, or any part of, the process involved in the mining of minerals on mined lands by removing overburden and mining directly from the mineral deposits, open-pit mining of minerals naturally exposed, mining by the augur method, dredging and quarrying, or surface work incident to an underground mine.[1]
>
> *Id.* § 2735

To determine if SMARA's reclamation requirements are applicable, the first subject to address is whether a person is engaged in "surface mining operations."

By this definition, a surface mining operation also includes excavation, crushing, and processing activities associated with mining as well as "[i]n place distillation or retorting or leaching"; the "production and disposal of mining waste"; and "[p]rospecting and exploratory activities." *Id.*

For SMARA to apply, however, the activities must relate to the mining of **minerals**. These include "[a]ny naturally occurring chemical element or compound, or groups of elements and compounds, formed from inorganic processes and organic substances, including, but not limited to, coal, peat, and bituminous rock...." Cal. Code Regs. tit. 14, § 3501. Readers with a geologic background may correctly note that common rock and gravel are sometimes not classified as minerals in a scientific sense. Real property law may also not treat those materials as minerals for purposes of reserving mineral rights. But this is not the case under SMARA, which includes rock, sand, and gravel as minerals subject to its requirements. *See* 66 Ops. Cal. Att'y Gen. 80 (1983).

To be subject to reclamation, surface mining operations must involve the mining of minerals that include common rock and gravels.

Mined lands. After determining that a person or entity is engaged in an activity or activities that constitute a surface mining operation, the extent of the **mined lands** must be established. SMARA defines this as:

Another subject to address in determining the applicability of reclamation is the extent of the operator's "mined lands."

> [T]he surface, subsurface, and ground water of an area in which surface mining operations will be, are being, or have been conducted, including private ways and roads appurtenant to any such area, land excavations, workings, mining waste,[[2]] and areas in which structures, facilities, equipment, machines, tools, or other materials or property which result from, or are used in, surface mining operations are located.
>
> Pub. Res. Code § 2729

Because of the breadth of this definition, almost any land used for or disturbed by a surface mining operation must be reclaimed. Moreover, SMARA provides that reclamation "may extend to affected lands surrounding mined lands," which makes clear that an operator may also have to reclaim non-mined property adversely affected by the mining operations. *Id.* § 2733.

"Mined lands" may include non-mined property adversely affected by mining operations.

SMARA also applies to mined lands on federal property. As explained in chapter 7, the United States Supreme Court's decision in *California Coastal Commission*

1. SMARA's regulations specify that surface mining also includes borrow pitting, streambed skimming, and stockpiling of mined materials. Cal. Code Regs. tit. 14, § 3501.

2. "Mining waste" is defined as "the residual of soil, rock, mineral, liquid, vegetation, equipment, machines, tools, or other materials or property directly resulting from, or displaced by, surface mining operations." Pub. Res. Code § 2730.

SMARA's reclamation requirements also apply to mining on federal land per the United States Supreme Court's Granite Rock *decision.*

v. Granite Rock Company (1987) 480 U.S. 572 has been interpreted to permit state and local environmental regulation on federal land so long as that regulation does not prohibit mining. Because SMARA mandates only how mined property must be reclaimed and not whether mining may occur, its requirements are not likely to conflict with federal mining laws. *See* 60 Ops. Cal. Att'y Gen. 162 (1977) (reaching the same conclusion prior to the *Granite Rock* decision). Additional support for this conclusion can be found in Public Resources Code § 2714(e), which exempts from SMARA mining operations conducted solely to protect a federal mining claim. This provision would be unnecessary were SMARA not applicable to federal property.

The final subject to address in determining the applicability of reclamation is who the mine "operators" are.

Operators. After determining whether surface mining operations are involved and the extent of the mined lands, the **operator** of the project must be established. SMARA defines an **operator** to include:

> [A]ny person who is engaged in surface mining operations, himself, or who contracts with others to conduct operations on his behalf, except a person who is engaged in surface mining operations as an employee with wages as his sole compensation.
>
> Pub. Res. Code § 2731

SMARA provides that operators include those who directly conduct the mining operations as well as those who contract with others to conduct mining on their behalf.

This definition clearly includes the persons or entities who directly conduct the surface mining operation. But by adding those "who contract[ ] with others to conduct operations on [their] behalf," the Act may arguably bind others who do not actually carry on the operation. For example, if contracting for mining makes one an operator, then the operator might also include a property owner who leases a mine site to a mining company for some form of profit, such as royalties. Because SMARA does not further define those "who contract with others," it is difficult to identify exactly who the "other" operators might be.

However, SMARA does make clear that persons or entities required to reclaim their sites includes those who operate under a claim of vested rights as well as those operating under permits. *See* chapter 8 (discussing vested or nonconforming mining uses). Person or entities who have operations that existed before SMARA's adoption in 1976 must also reclaim their mine sites, but only for lands disturbed after SMARA's effective date of January 1, 1976. Cal. Code Regs. tit. 14, § 3505(b). Any operator who continues to mine in or on an area disturbed before the Act's adoption—creating a vertical or horizontal overlap between pre- and post-SMARA mining—must only reclaim an area proportionate to the post-SMARA disturbance. *Id.*

Persons or entities with operations that pre-existed SMARA's adoption must reclaim their operations, but only in proportion to the land disturbed since the Act's effective date.

Once it is approved, a reclamation plan is binding on all successor operators. Pub. Res. Code § 2779. Moreover, because SMARA requires that a reclamation plan identify the name, address, and registered agent for the operator, the plan should be amended to reflect the new operator information upon transfer. *See id.* § 2772(c)(1). The amendment must also incorporate the operator's statement accepting responsibility for reclaiming the operation. *Id.* § 2772(c)(10).

Activities Exempt from Reclamation

Certain activities that would otherwise constitute surface mining operations are exempt from SMARA's reclamation requirements. After considering the provisions discussed above, one must determine if any of the following exemptions apply.

Construction-related excavation. SMARA exempts excavation and earthmoving activities that are "an integral and necessary part" of a construction project. Pub. Res. Code § 2714(b). For purposes of this exemption, a construction project includes activities needed to prepare a site for "construction of structures, landscaping, or other land improvements associated with those structures, including the related excavation, grading, compaction, or the creation of fills, road cuts, and embankments...." *Id.* The exemption applies, however, only if the project is consistent with the lead agency's general plan and conforms to the zoning of the site. The lead agency must also have issued all necessary entitlements for the construction and analyzed its effects per the California Environmental Quality Act. *Id.* § 2714(b)(1)–(3).

SMARA exempts from its reclamation provisions excavation and earthmoving activities that are an integral and necessary part of a construction project.

If the construction exemption applies, the contractor may transport surplus excavated materials offsite, but may only do so during the actual construction of the project site. *Id.* § 2714(b)(4). Once construction is completed, or once it has been suspended or terminated, the operator must cease offsite transportation of material.

Prospecting and extraction of less than 1,000 cubic yards of material. Prospecting that disturbs fewer than 1,000 cubic yards in any location of one acre or less is not subject to SMARA. Pub. Res. Code § 2714(d). Although SMARA's definition of a "surface mining operation" literally includes prospecting (Pub. Res. Code § 2735(c)), this exemption excepts most prospecting from reclamation.

Prospecting that disturbs fewer than 1,000 cubic yards is not subject to reclamation.

An excavation of fewer than 1,000 cubic yards of materials, ores, and overburden—or an excavation involving less than one acre of disturbance—are likewise exempt. Pub. Res. Code § 2714(d); Cal. Code Regs. tit. 14, § 3505(a)(1). This exemption excludes common and minor excavations for landscaping, irrigation, drainage, and household purposes that would literally—and oddly—be subject to reclamation were there no such exemption.

Mineral processing. An operation that strictly processes mined material—including recycled concrete and asphalt—is not required to reclaim the site. Pub. Res. Code § 2714(c). This exemption does not apply when an operator conducts any onsite excavation. *Id.* § 2714(c)(3). The exemption also requires that the processing operation occur on property designated in the lead agency general plan and applicable zoning district for commercial or industrial use. *Id.* § 2714(c)(1)(2).

Operations that strictly process mined materials are not subject to reclamation.

Timber operations and forest management. Excavation or grading for roadbed construction and maintenance for timbering operations or forest management do not require reclamation. Pub. Res. Code § 2714(j)(1). But this exception is limited to lands adjacent to timber operations or forest management roads and does not apply

Excavations or grading for roadbed construction and maintenance for timbering operations or forest management need not be reclaimed.

to excavation located within a certain distance of Class I or Class II watercourses. *Id.* The exception also does not apply if the excavated material is commercially sold. *Id.* Any person claiming the exemption must further meet prescribed slope and erosion requirements and must revegetate disturbed areas where necessary. *Id.* § 2714(j)(2).

Excavation, grading, and earthmoving activities in oil and gas fields that are integral to and necessary for extraction are exempt from SMARA.

Oil and gas operations. Excavation, grading, and earthmoving activities in oil and gas fields are exempt from SMARA if "integral to, and necessary for" ongoing or gas extraction. *Id.* § 2714(k). The exception only applies when excavated materials are not sold, the oil and gas field properties are under a common owner, and operations are consistent with applicable general plan and zoning requirements. *Id.* § 2715(k).

Solar evaporation of sea and bay water is not subject to SMARA.

Solar evaporation of sea or bay water. Solar evaporation of sea and bay water for the production of salt and related minerals is not subject to SMARA. *Id.* § 2714(g).

Mining conducted exclusively to protect a federal mining claim is exempt from SMARA.

Mining operations required by federal law to protect mining claim. A surface mining operation conducted exclusively to protect a federal mining claim is exempt from California's reclamation requirements. *Id.* § 2714(e). However, they may be subject to reclamation under federal law.

Excavation and grading necessary for farming or to restore land after a flood or natural disaster are not subject to reclamation.

Land disturbance for farming or restoration after a natural disaster. Excavation or grading necessary for farming activities is not subject to SMARA. *Id.* § 2714(a). Excavation and grading necessary to restore land after a flood or natural disaster are likewise exempt. *Id.* An excavation for agricultural activities involving the extraction of more than 1,000 cubic yards of material for commercial purposes, however, is subject to SMARA. Cal. Code Regs. tit. 14, § 3505(a)(3).

SMARA is not applicable to Department of Water Resources activity to alleviate, repair, or restore damage to property due to floods, disasters, or emergencies.

Emergency excavations by the Department of Water Resources. SMARA does not apply to any Department of Water Resources activities for alleviating, repairing, or restoring damage to property due to floods, disasters, or other emergencies. Pub. Res. Code § 2714(h). The exemption applies both to "recent" emergencies and those that are "imminent."

Mining on lands the Department of Water Resources or Reclamation Board owns, leases, or for which there is an easement is not subject to SMARA.

Mining for or by the Department of Water Resources and Reclamation Board. Mining on lands owned or leased by the Department of Water Resources or Reclamation Board, or on lands on which those agencies have easements or rights of way, is not subject to SMARA's reclamation requirements if done for the State Water Resources Development System or for purposes of flood control. Pub. Res. Code § 2714(i)(1). To invoke this exemption, the Department of Water Resources must adopt its own regulations setting forth the reclamation requirements for such sites in consultation with the Department of Conservation. *Id.* Any operator carrying out an agency project must then comply with the agency's reclamation regulations.

SMARA empowers the SMGB to adopt additional exemptions.

Other exemptions. SMARA empowers the SMGB to create exemptions for activities the SMGB "determines to be of an infrequent nature and which involve only minor surface disturbances." Pub. Res. Code § 2714(f). Since these change from time to time, the reader is advised to consult the most recent version of SMGB-adopted regulations. As this book goes to press, the administrative exemptions can be found at California Code Regulations, Title 14, section 3505.

The Reclamation Plan

An entity that qualifies as an operator of a surface mining operation under the standards described above must receive lead agency approval of a reclamation plan. The plan must also cover all of the operator's mined lands. After approval, the plan remains in effect for the lifetime of the operation and, following conclusion of mining, until reclamation is complete.

Operators of surface mining operations must receive approval of reclamation plans covering the entirety of their mined lands.

The plan's volume and complexity, which typically draws on geology, geography, biology, engineering, and other scientific disciplines, can be overwhelming. However, its purpose is simple, and no science background is necessary to understand why it is required. At the most basic level, a reclamation plan is essentially a contract with a lead agency affirming that the operator will restore the land to productive use when mining is complete. It is also a detailed instruction manual that governs how an operator must fulfill its obligations.

Reclamation plans may be voluminous and complex, but their basic purposes are easily understood—they affirm that operators will reclaim their properties following mining and provide a detailed plan for how this will be done.

General Requirements for Reclamation Plans

Every person or entity who operates a surface mining operation must receive approval of a reclamation plan. Pub. Res. Code § 2770(a). The plan must further apply to "a specific piece of property or properties." *Id.* § 2773(a). Operators therefore must obtain reclamation plans for each of their operations; they cannot have a single plan covering multiple mining sites. Nor can they have multiple plans for the same mining site.[3]

A mining operator must obtain approval of a reclamation plan and may not obtain more than one plan for an operation.

For each individual operation, the reclamation plan must be developed to attain SMARA's legislative policies. Cal. Code Regs. tit. 14, § 3502(a). The thrust of the plan thus must be to "[restore the] mined land... to a useable condition which is readily adaptable for alternative land uses," to "minimize adverse effects on the environment," and to "protect the public health and safety." Pub. Res. Code §§ 2711(a), 2712(a). In implementing these policies, a reclamation plan must also be tailored to suit the project's nature and surroundings. As SMARA provides, a plan must be based on "the character of the surrounding area and such characteristics of the property as type of overburden, soil stability, topography, geology, climate, stream characteristics, and principal mineral commodities...." *Id.* § 2773(a). A reclamation plan must also take into account "physical and land-use characteristics of the mined lands and their surrounding[s]...." Cal. Code Regs. tit. 14, § 3502(c).

Every reclamation plan must be designed to attain SMARA's legislative policies.

3. The only exception to this requirement is when a single surface mining operation has facilities located in more than one jurisdiction and those facilities are separated by a distinct physical boundary such as a highway or stream channel. Cal. Code Regs. tit. 14, § 3502(i)(1). In that situation, the operator may obtain separate reclamation plans from all the lead agencies having jurisdiction over the mine site. *Id.*

Every reclamation plan must establish site-specific criteria for evaluating operator compliance.

Each plan must develop site-specific criteria for evaluating compliance. Pub. Res. Code § 2773(a). For most operations today, these criteria are based on minimum, verifiable standards set forth in regulations adopted by the SMGB. *Id.* § 2773(b). The performance standards apply to any new mining operation, any operation whose plan was adopted in 1993 or later, or any operation that has substantially amended a reclamation plan approved before 1993. Cal. Code Regs. tit. 14, § 3700(c)–(d). However, the standards apply only to the extent they are consistent with the "planned or actual subsequent use or uses of the mining site." Pub. Res. Code § 2773(b). A mining operation that would not disturb wildlife habitat or be converted to agricultural use, for example, would not be subject to performance standards for wildlife habitat or agricultural land reclamation.

Cities and counties may allow operators to deviate from an SMGB-imposed performance standard only if necessary and only if the alternative standard is no less stringent.

A city or county may allow an operator to deviate from one or more SMGB-established performance standards, but only if the operator demonstrates that an exception to a standard "is necessary based upon the approved end use." Cal. Code Regs. tit. 14, § 3700(b). In allowing any such exception, the lead agency must establish its own verifiable, site-specific reclamation standards that are no less stringent than the SMGB-prescribed standard. *Id.*

Reclamation Performance Standards

The SMGB has promulgated several performance standards for reclamation. Because the standards are complex and science-based, those involved in plan preparation or review usually need the assistance of biologists, geologists, engineers, and other professionals for interpretation and compliance.

Recontouring. By its nature, mining results in a large amount of land disturbance, often resulting in deep quarry pits or acres-wide working areas. Converting mined-out property to land that can subsequently support a productive use usually requires substantial earth movement, filling, grading, and reshaping. These activities can be the most time-consuming and costly aspects of reclamation. In addition, SMARA's performance standards for recontouring differ for a metallic or non-metallic mining operation, and special recontouring standards apply for an operation on or near certain Native American sites.

The standards for recontouring depend on whether the operation mines metallic or non-metallic minerals.

Recontouring of non-metallic mines. Hard rock, sand and gravel, and other non-metallic operations must meet minimum standards for recontouring their properties. For pits, working areas, and other depressions where backfilling is required, an operator must meet soil compaction standards appropriate for the mining site's end use. Cal. Code Regs. tit. 14, § 3704(a), (b). If a proposed end use involves an urban use, for instance, the backfill material must be compacted in accordance with the Uniform Building Code or other specific standards the lead agency has adopted. *Id.* § 3704(a).

If backfilling is required, the fill material must meet soil compaction standards appropriate for the operation's end use.

Much of the material an operator uses for backfill comes from overburden and waste piles generated during a mining operation. In this situation, the performance standard requires that the material be stockpiled in a manner to facilitate phased

Material stored for backfilling must be stockpiled in such a manner as to facilitate phased reclamation.

reclamation. *Id.* § 3704(c). When the overburden and waste piles are to become a permanent feature of the land, the slopes of those piles—called **fill slopes**—must conform to the surrounding topography or the topography of the end use. *Id.* § 3704(e). To ensure stability, the slopes of the fill piles must not exceed a 2:1 horizontal-to-vertical slope, unless a geologic or engineering analysis demonstrates that a different slope is feasible. *Id.* § 3704(d).

Fill slopes must conform topographically with surroundings and usually cannot have slopes exceeding a 2:1 ratio.

In some cases, an operator may choose to have highwalls and a working face become a feature of the property's subsequent use. For residential development, for example, creating tiered levels or using a mine wall as a scenic backdrop may greatly enhance the site's second-use value. If such a **cut slope** remains, at a minimum the stability factor must be suitable for the proposed end use. *Id.* § 3704(f). Slopes must also conform to the surrounding topography of the proposed use. *Id.*

Cut slopes must conform with their surroundings and have slope stabilities suitable for their end uses.

Recontouring of metallic mines. In 2002, the SMGB promulgated special backfilling standards for metallic mines. Initially adopted as an emergency measure, the standards addressed concerns that reclamation plans for metallic mines were permitting the retention of large pits, overburden, and waste piles as part of a post-mining land use. The mine operators had characterized such reclaimed conditions as open space. However, the SMGB did not agree and adopted a new regulation requiring that all open pit excavations be backfilled to at least the area's original surface elevation. Cal. Code Regs. tit. 14, § 3704.1(a). The regulation further requires that an operator use waste and overburden piles that must first be treated, if necessary, to remove contaminants. *Id.* § 3704.1(b).

Metallic mines generally must backfill open-pit excavations to their original contours.

Under the new standards, an operation is considered a metallic mine if more than 10 percent of its gross revenue, averaged over the previous five years, was derived from the production of any or a combination of certain metallic minerals[4] by the open-pit extraction method.[5] An operator is only exempt from the standards when all overburden and waste piles remaining at the conclusion of mining operations are insufficient in volume to completely backfill the open-pit excavation to its pre-mining surface. *Id.* § 3704.1(h). In that situation, an operator need only backfill to an elevation that utilizes all of the material remaining as mining waste or overburden.

If overburden and waste piles are insufficient in volume to completely backfill excavations, metallic operators need only backfill to an elevation that utilizes all the overburden and mining waste.

The resulting topography from the backfilling required by this standard must not exceed the pre-mining contours by more than 25 feet. *Id.* § 3704.1(e). Final fill slopes must also meet minimum engineering requirements, be capable of revegetation, and blend in with surrounding topography. *Id.* Fills and fill slopes must be

4. These include precious metals (gold, silver, platinum), iron, nickel, copper, lead, tin, ferro-alloy metals (tungsten, chromium, manganese), mercury, uranium and thorium, minor metals (such as strontium, rubidium, cesium), and niobium and tantalum. Cal. Code Regs. tit. 14, § 3704.1(f).

5. Per this regulation, an open-pit mine "is the same as an open pit quarry, opencast mine, or opencut mine and is defined as a mine working or excavation that is open to the surface and in which the opening is approximately the full size of the excavation." Cal. Code Regs. tit. 14, § 3704.1(g).

further designed to protect groundwater quality, prevent surface water ponding, facilitate revegetation, convey runoff in a non-erosive manner, and account for long-term settlement. *Id.* § 3704.1(d).

Special recontouring standards also apply to metallic mining operations on or near Native American sacred sites.

Recontouring of metallic mines on or near Native American sacred sites and areas of concern. Special metallic mining standards also apply to operations on or near certain Native American sacred sites. In 2002, the legislature amended SMARA to protect those sites "from the potential environmental impacts and depredations associated with cyanide heap leach mining, which leaves huge holes in the ground and extremely large mounds of leached ore materials and mining wastes above ground." Assembly Committee on Natural Resources, Analysis of SB 483, page 4 (Aug. 29, 2002).

Under the 2002 legislation, a lead agency may not approve a reclamation plan for gold, silver, copper, or other metallic mines located on, or within one mile of, a Native American sacred site if that land is an area of special concern, unless the operator meets certain backfilling and grading criteria. Pub. Res. Code § 2773.3(a); *see also id.* § 2773.3(b)(1) (defining "Native American sacred site"). Areas of special concern are those designated as Class C or Class L lands in the California Desert Conservation Area. *Id.* § 2773.3(b)(2); *see also* 43 U.S.C. § 1781. When subject to the legislation, a reclamation plan must require that all excavations be graded and backfilled to achieve the original contours of the mined land before grading. Pub. Res. Code § 2773.3(a)(1)(A). Further, overburden, spoil piles, and heap leach piles in excess of the materials that can be placed back into the excavated area must be graded over the remainder of the project site to achieve the original contours of the mined land. *Id.* § 2773.3(a)(1)(B).

Operations on or near Native American sacred sites must backfill excavations to original contours.

Revegetation. In addition to requiring that mined land be recontoured, a reclamation plan typically requires that the land be planted with vegetation suitable for the proposed end use and for controlling erosion. Called **revegetation,** this process, like recontouring, is a common feature of a reclamation plan. SMARA's revegetation performance standard generally mandates that mined land be planted with "a vegetative cover suitable for the proposed end use and capable of self-regeneration without continued dependence on irrigation, soil amendments, or fertilizer. . . ." Cal. Code Regs. tit. 14, 3705(a). Revegetation must further be sufficient to "stabilize the surface against effects of long-term erosion and shall be similar to naturally occurring habitats in the surrounding area." *Id.*

Mined land must be replanted with species suitable for the lands' end uses, be capable of self-regeneration, and prevent erosion.

Ordinarily, an operator must revegetate with native plant species. *Id.* § 3705(g); *see also id.* § 3701 (defining "native species" as plant species "indigenous to California, using pre-European as the historic time reference."). But an operator may use non-native plants if they are necessary for the reclamation plan's proposed end use. *Id.* § 3705(g). Industrial, commercial, or residential end uses also need not use native vegetation, so long as the vegetation used does not involve noxious weeds. *Id.*; *see also id.* § 3701 (defining **noxious weeds**). Whether using native or non-native vegetation, the operator must employ revegetation methods and plant species that,

Revegetation must usually be accomplished with native plant species, unless the mined lands' end uses require otherwise.

according to available research, has good survival properties for the topography, its resoiling characteristics, and the climate involved. *Id.* § 3503(g).

Identifying the specific seed mixes, soils, and conditions necessary to fully revegetate a mined site—especially one on which mining has occurred for decades—is not an easy task. Because revegetation is done by trial and error, this performance standard generally requires that an operator conduct test plots during mining to "determine the most appropriate planting procedures to be followed to ensure successful implementation of the proposed revegetation plan." *Id.* § 3705(b). After beginning the revegetation process following mining, the operator must also take appropriate measures to protect the revegetating areas. This may involve placement of cages or fencing to prevent grazing, trampling, or other attacks from herbivores. *Id.* § 3705(l).

Operators must conduct test plots during mining to ensure successful implementation of revegetation.

To demonstrate full revegetation, the operator must ultimately meet quantified standards set forth in the reclamation plan. *Id.* § 3705(m). Overall, success is "judged based upon the effectiveness of the vegetation for the approved end use, and by comparing the quantified measures of vegetative cover, density, and species-richness of the reclaimed mined-lands to similar parameters of naturally occurring vegetation in the area." *Id.*

Topsoil salvage. Revegetation usually requires large amounts of soils and other growth media for native plants and other vegetation to be planted. An additional performance standard therefore requires that an operator remove and store all salvageable topsoil from areas to be disturbed by the mining operation. Cal. Code Regs. tit. 14, § 3711(a); *see also id.* § 3501 (defining **topsoil**). If the salvaged topsoil will not cover all of the surfaces to be revegetated as part of reclamation, an operator may salvage and store other suitable material capable of sustaining vegetation, such as subsoil. *Id.* § 3711(b).

To assist in revegetation, SMARA requires that operators salvage topsoil from disturbed areas.

All stockpiles of topsoil and growth media must be planted with a vegetative cover and must be protected by other measures that discourage weeds and prevent water and wind erosion. *Id.* § 3711(d). During the mining phase preceding reclamation, an operator must also store topsoil and other growth media in piles separate from each other and from other piles, such as those containing mining waste. *Id.* § 3711(b), (d). An operator must further store the piles of topsoil and growth media in locations that will not be disturbed prior to reclamation. *Id.* § 3711(d). Relocation of the piles requires approval by the city or county. *Id.*

Operators must plant topsoil piles with a vegetative cover, must store such piles separately from mining waste and overburden piles, and cannot move the piles without permission.

To ensure full performance under the salvaging performance standard, a reclamation plan must provide a map of all topsoil resources and the location of all soil stockpiles. *Id.* § 3711(b). The reclamation plan must further identify a schedule for salvaging soil related to the phases of reclamation that both minimizes the area of soil disturbance and achieves the maximum revegetation success allowable under the mining plan. *Id.* § 3711(c). Once reclamation begins, the topsoil and other salvage material must be distributed through the areas to be revegetated "in a manner that results in a stable, uniform thickness consistent with the approved end use, site configuration, and drainage patterns." *Id.* § 3711(e).

The reclamation plan must identify the location of the topsoil piles and provide a soil-salvaging schedule.

Building, structure, and equipment removal. Mining operations require buildings, facilities, and structures to house plants, equipment, supplies, and other necessities. As part of reclamation, these must be removed after completion of mining, unless the reclamation plan specifies that some or all remain as part of the post-mining use. Cal. Code Regs. tit. 14, § 3709(b). Onsite post-mining waste must also be disposed of in accordance with state and local health ordinances. *Id.* § 3709(a).

SMARA requires that operators comply with a number of requirements to control erosion, restore drainage patterns, and maintain water quality.

Drainage, erosion, and water resources. An additional performance standard is designed to remediate adverse mining impacts associated with erosion and alteration of drainage patterns and water resources. Under this standard, an operator must control erosion and sedimentation to minimize siltation of lakes and water courses as required by the Regional Water Quality Board or State Water Resources Control Board. Cal. Code Regs. tit. 14, § 3706(c). An operator must also protect other water bodies and surrounding lands from erosion, gullying, sedimentation, and contamination by using berms, silt fences, sediment ponds, revegetation, hay bales, or similar measures. *Id.* § 3706(d). Onsite and downstream beneficial uses of the water must also be protected per the Porter-Cologne Water Quality Control Act and federal Clean Water Act. *Id.* § 3706(a); *see also* Water Code §§ 13000 *et seq.;* 33 U.S.C. §§ 1251 *et seq.* Likewise, surface and groundwater must be protected from siltations and pollutants that may diminish water quality. Cal. Code Regs. tit. 14, § 3710(a).

A mining operation must not diminish the quality of water, recharge potential, and storage capacity of groundwater aquifers, except as provided in an approved reclamation plan. *Id.* § 3706(b). Whenever a surface mining operation covers, reroutes, or has an impact on a natural drainage, the reclamation plan must require measures ensuring that runoff shall not cause increased erosion of sedimentation. *Id.* § 3706(e). Any required stream diversion must be constructed in accordance with a stream and lake alteration agreement between the operator and the Department of Fish and Game and other applicable federal laws. *Id.* § 3706(f); *see also* 33 U.S.C. §§ 403, 1311, 1344.

Operators must prepare a post-closure maintenance plan to ensure that they do not degrade water quality.

Tailing and mine waste management. In reclaiming mine waste to remain after mining is complete, an operator must comply with standards prescribed by the State Water Resources Control Board. Cal. Code Regs. tit. 14, § 3712; *see generally* Cal. Code Regs. tit. 27, §§ 22470–22510. The provisions generally require a post-closure maintenance plan guaranteeing that new and existing mine units are closed "so that they no longer pose a threat to water quality." *See* Cal. Code Regs. tit. 27, § 22510(a)–(b). Specific closure standards apply depending upon the type of waste and its potential to contain hazardous substances. *Id.* § 22510(j)–(l). The plan must also ensure that waste containment, precipitation, and drainage control standards are met during the mine's closure and post-closure periods. *Id.* § 22510(b); *see also id.* § 22510(h) (indicating that the post-closure period extends until the regional water quality control board "determines that water quality aspects of reclamation are complete and waste no longer poses a threat to water quality"). The operator must also post a financial assurance to guarantee performance, which can be incorporated into the SMARA financial assurance if the operator names the regional water quality control board as an additional payee." *Id.* § 22510(g). Waste discharge requirements issued by a regional water quality control board may also incorporate the relevant provisions of the reclamation plan and impose additional requirements to control water degradation. *Id.* § 22510(c).

Closure of surface openings. Another SMARA performance standard mandates that, prior to closure of a mining site, an operator must gate or protect from public entry all portals, shafts, tunnels, and other surface openings. Cal. Code Regs. tit. 14, § 3713(b). The primary purpose of this requirement is to prevent threats to public safety and, secondarily, to protect wildlife habitat.

SMARA requires that operators close all mine portals, shafts, tunnels, and other surface openings.

Upon reclamation, all drill holes, water wells, and monitoring wells must be completed or abandoned per various water laws. *Id.* § 3713(a). Generally found in California Water Code §§ 13700 *et seq.* and 13800 *et seq.*, these provisions govern abandonment of all types of wells and require that local agencies adopt well abandonment ordinances. Any applicable local ordinances should be consulted for additional requirements.

Wildlife habitat. Mining often occurs on lands inhabited by sensitive plants and animal species. As discussed in chapter 13, state and federal endangered species laws already regulate the extent to which a mining operation can disturb species and habitat along with the kind of mitigation required as a result of disturbance. SMARA's performance standards address those matters further in relation to post-mining reclamation.

The habitat performance standards for wildlife first incorporate by reference all federal and California Endangered Species Act requirements. The standards thus provide that reclamation activities conserve rare, threatened, and endangered species and their habitat unless the operator provides appropriate mitigation for the disturbance. Cal. Code Regs. tit. 14, § 3703(a); *see also* 16 U.S.C. §§ 1531 *et seq.*; Fish & Game Code §§ 2050 *et seq.* In addition, disturbed habitat must be reclaimed in a condition comparable to what existed before mining, unless the end use of the site either precludes its use for wildlife or the reclamation plan establishes a different type of habitat than previously existed. Cal. Code Regs. tit. 14, § 3703(b).

In addition to complying with federal and state endangered species laws, operators must restore habitat disturbed by mining to as good a condition as existed before mining.

Reclamation activities must also avoid any disturbance to wetland habitat. *Id.* § 3703(c); *see also id.* § 3701 (adopting the "wetlands" definition of Fish & Game Code § 2785(g)). Any wetland habitat disturbed by a mining operation must be mitigated at a minimum 1:1 ratio for wetland habitat acreage and wetland habitat value. Cal. Code. Regs. tit. 14, §3703(c); *see also id.* § 3704(g) (prohibiting placement of permanent overburden or overburden piles on wetlands unless the appropriate agencies approve mitigation to offset wetlands losses).

Operators must restore any wetlands disturbed by mining to a 1:1 ratio for acreage and habitat value.

In-stream and river-channel operations. A mining operation that extracts materials deposited by streams or rivers must comply with the federal Clean Water Act, Porter-Cologne Act, county anti-siltation ordinances, and any regulations of the

In-stream mining operations must comply with requirements to protect water quality and aquatic species.

Regional Water Quality Control Board or the State Water Resources Control Board. Cal. Code Regs. tit. 14, § 3710(b). The operation must not cause fish to become entrapped in pools or off-channel pits and may not restrict spawning or migratory activities. *Id.* § 3710(d).

An operator who extracts material from a river channel must control degradation. Cal. Code Regs. tit. 14, § 3710(c). The mining operation must also prevent the undermining of bridge supports, exposure of pipelines or other structures buried in the channel, loss of spawning, habitat, lowering of groundwater levels, destruction of vegetation, and increased stream bank erosion. *Id.* All changes to channel elevation and bank erosion must be evaluated annually. *Id.*

Land reclaimed as prime farmland must for two years be equivalent to or exceed the pre-mining condition of land or similar crop production in the area.

Agricultural land reclamation. Land reclaimed for agricultural use must meet certain standards for crop viability, and any land that must be converted to prime agricultural land shall have layered soil horizons in accordance with U.S. Soil Conservation Service standards. Cal. Code Regs. tit. 14, § 3707(b). The reclamation plan must also specify the appropriate fertility level for the reclaimed soil. *Id.* § 3707(a). To be complete, for two crop years reclamation of the land must be equivalent to, or exceed, the land's pre-mining condition or similar crop production in the area. *Id.* § 3707(c).

Land reclaimed to non-prime farmland must be capable of sustaining economically-viable production of crops grown in surrounding areas.

Land not reclaimed to prime agricultural land must, at a minimum, be capable of sustaining economically viable production of crops commonly grown in the areas surrounding the mining operation. *Id.* § 3708.

Phasing of Reclamation

In recent years, reclamation plans have begun to emphasize a process called **phasing**, the interim reclamation of certain areas before terminating the entire operation. At any one time during the life of a surface mine, extraction will be concentrated in a particular area. Consequently, that area's resources may be depleted many years before final closure of the mine. Instead of allowing a mined-out area to remain, many plans now require that the operation be reclaimed in phases.

If achievable, phased reclamation is usually preferable to reclaiming all at once. Much of the reclamation can occur before mine closure, reducing the time between closure and implementation of the post-mining end use. With phasing an operator can also test the success of reclamation measures—particularly those pertaining to revegetation—long before mining terminates. This period of trial and error can early on reveal any flaws that need correction.

SMARA encourages phased reclamation, which allows operators to test the success of their reclamation measures well before mining concludes.

SMARA has a clear preference for phased reclamation. In describing the contents of a reclamation plan, discussed more fully below, the Act requires that an operator provide "a time schedule that will provide for the completion of surface mining on each segment of the mined lands so that reclamation can be initiated at the earliest possible time on those portions of the mined lands that will not be subject to further disturbance by the surface mining operation." Pub. Res. Code § 2772(c)(6). Generally, therefore, a reclamation plan should incorporate phasing.

Reclamation Plan Approval Process

Lead agency ordinance. Before an operator applies for reclamation plan approval, the lead agency must adopt a surface mining ordinance. Pub. Res. Code § 2774(a). Requirements for a lead agency mining ordinance are discussed in detail in chapter 6. However, it is important to note that the surface mining ordinance is a necessary precursor to processing an application for a reclamation plan, since the lead agency establishes procedures for the plan's review and approval and must implement state reclamation policy.

Before cities and counties can process reclamation plan applications, they must adopt a special mining ordinance.

Form and content of reclamation plan applications. An operator must submit a proposed reclamation plan to a lead agency "on...form[s] provided by the lead agency." Pub. Res. Code § 2772(a). Because the agency may not have all the forms at hand, a complete draft of the plan should be submitted. If other entitlements are needed, such as those for a conditional use permit or a change to the general plan and zoning, the reclamation plan should be submitted along with those requests. The reclamation plan need not duplicate information already stated in those applications if the plan incorporates the other requests by reference. *Id.* § 2772(d). However, the operator must submit all required documentation to the lead agency at one time. *Id.* § 2772(b).

Operators must usually apply for reclamation plans by submitting complete drafts of their plans to their cities or counties.

A reclamation plan must include all of the following:

- Name and address of the surface mining operator and the names and addresses of anyone the operator designates to accept service in its name
- Anticipated quantity and type of minerals the operator will mine
- Proposed dates of initiation and termination for the operation
- Maximum anticipated depth of the operation
- Geologic, geographic, topographic, and property information; specifically, the operators must identify the size and legal description of lands affected, provide a map that includes the boundaries and topographic details of the lands, and give a detailed description of the geology of the area where mining will occur; the operator must also identify the location of all streams, roads, railroads, and utility features on or adjacent to the land to be mined, and must further identify all proposed access roads to be constructed in conducting the operations and the names and addresses of all surface and mineral owners of the lands
- Timing of mining phases; the operator must describe and provide a plan for the type of mining to be employed, and must also provide a schedule that will allow mining of each phase to terminate as quickly as possible so that reclamation can begin on lands that will not be disturbed in the future
- Description of the proposed or potential end use or uses for the mineral lands and evidence that all owners of a possessory interest in the land are aware of those uses
- Description of how reclamation will be accomplished that explains how contaminants will be controlled and how mining waste will be disposed; if applicable,

the plan must additionally describe how streambed channels will be rehabilitated

- Assessment of the effect of implementation of the reclamation plan on future mining in the area
- Statement that the person submitting the reclamation plan accepts responsibility for reclaiming the mined lands in accordance with the plan
- Any other information the lead agency may require by ordinance

Id. § 2772(c)

The SMARA regulations require additional information and documents as part of an application for a reclamation plan, including:

- Environmental setting of the site and effect that possible alternate reclaimed site conditions may have upon existing and future uses of surrounding lands
- Public health and safety concerns, with consideration of the degree and type of present and future exposure of the public to the site
- Slope and stability information; the designed steepness and proposed treatment of the mined lands' final slopes must consider physical properties of the slope material, its probable maximum water content, landscaping requirements, and other factors; every plan must specify the critical gradient for the type of material involved; if final slopes will approach the critical gradient, the operator must provide an engineering analysis of slope stability
- Identification of areas mined to produce additional materials for backfilling and grading, as well as settlement of filled areas; if the end use for the site includes roads, building sites, or other uses sensitive to settlement, the reclamation plan shall include compaction of fill materials in conformance with good engineering practice
- Disposition of old equipment
- Temporary steam or watershed diversions

Cal. Code Regs. tit. 14, § 3502(b)(1)–(6)

When new mining operations are proposed, cities and counties process reclamation plan applications simultaneously with applications for conditional use permits and other entitlements.

Hearing and approval process for reclamation plans. In some cases, a mining operator may apply for a reclamation plan or a plan amendment without seeking additional entitlements. When a new or expanded mining use is proposed, however, an operator often applies for a reclamation plan in conjunction with other requests—for example, a general plan amendment, a zoning change, or a conditional use permit.

When this occurs, the hearing process for approval of a reclamation plan occurs simultaneously with the approval process for any other required entitlements. The process for reclamation plan approval described below therefore overlaps significantly with the one used for land use entitlements, which will be discussed in chapter 7. Figure 5-1 shows the approval process graphically, highlighting steps unique to reclamation.

The specific department assigned to review a reclamation plan varies from agency to agency. Often, a planning or community development department will

Figure 5-1

Reclamation Plan and Financial Assurance Approval Process

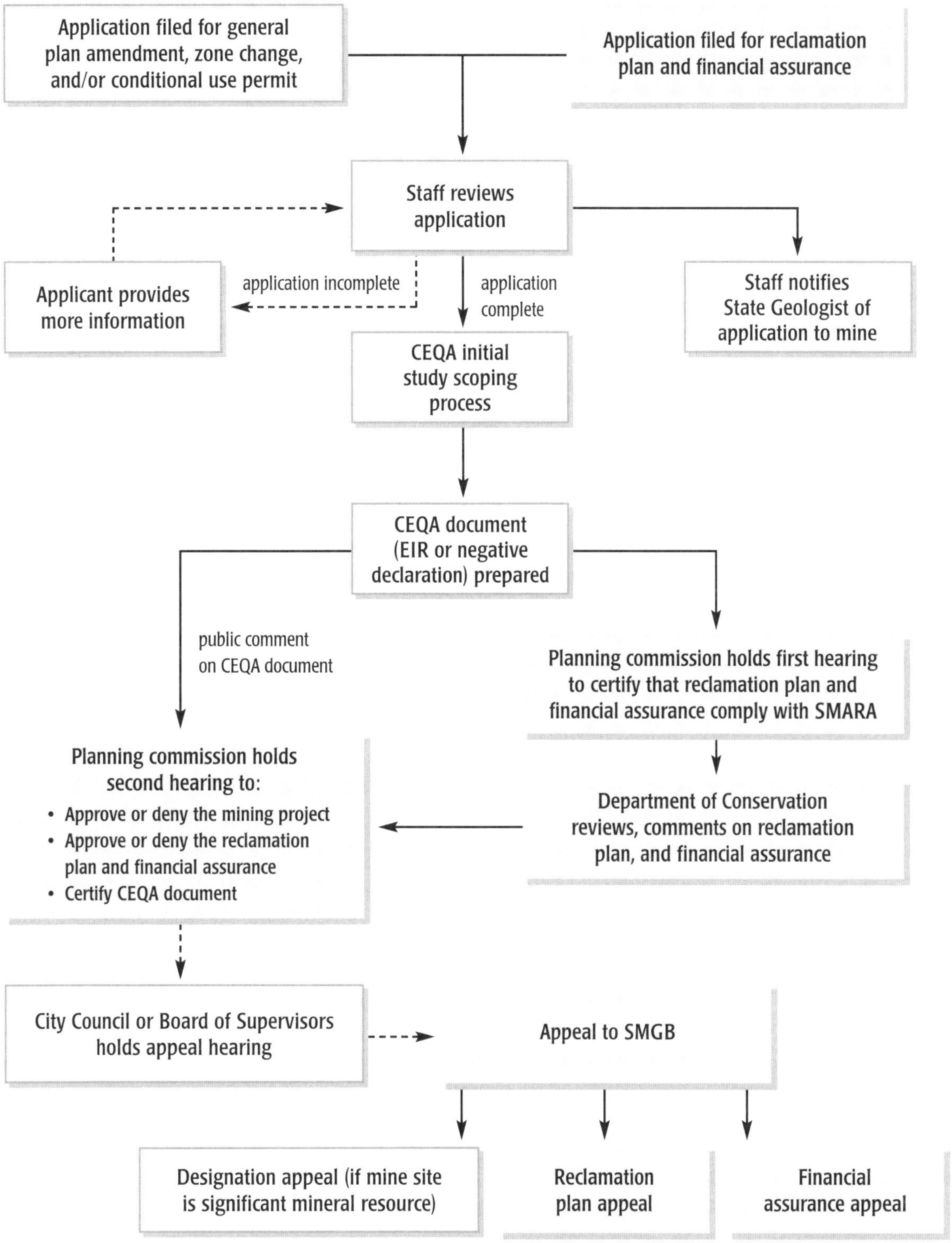

When planning or public works departments review reclamation plans, their first task is to determine that applications are complete.

review the plan, while another agency might assign review to public works or engineering. However the agency handles the review, the first task is to have lead agency staff determine that the proposed plan is complete. To make that determination, staff should be certain that the operator has provided every document and all information described above and that the plan meets all performance standards described previously in this chapter. Staff may need to retain consultants to assist in the effort.

At the beginning of the application process, staff must prepare an initial study per CEQA to determine the reclamation plan's potential environmental impacts.

CEQA = California Environmental Quality Act

After determining that the submission is complete, staff must conduct an initial study under the California Environmental Quality Control Act (CEQA) to determine whether the proposed plan creates the possibility of any significant and adverse impacts on the environment. *See* chapter 8 (providing a more thorough discussion of CEQA mining project review). Typically, a reclamation plan by itself would not pose a potential adverse impact since its purpose is beneficial to the environment—specifically, to restore and rehabilitate mined land. Thus, only a negative declaration may be required. Whatever the decision, though, staff must ensure that the appropriate document is circulated for public review and comment according to CEQA.

The city or county planning commission must determine if the reclamation plan substantially meets SMARA's standards.

Once staff determines that the reclamation plan is complete and the appropriate CEQA document has been prepared, the lead agency must hold at least one public hearing. Pub. Res. Code § 2774(a). This is usually conducted by the planning commission. After following standard hearing procedures, the commission must determine whether the proposed reclamation plan "substantially meet[s]" all SMARA requirements or regulations and any additional requirements under the lead agency's surface mining ordinance. *Id.* § 2770(d). If those requirements are not met, the reclamation plan must be returned to the operator, who then has 60 days to resubmit a plan correcting the plan's deficiencies. *Id.*

The city's or county's function in reviewing a reclamation plan is technical; it determines only whether the application substantially meets the requirements of SMARA.

If the proposed plan "substantially meets" all SMARA and lead agency requirements, the lead agency *must* approve the plan. Unlike a conditional use permit or similar entitlements, which the agency has discretion to approve or deny, a lead agency's power to deny a reclamation plan is limited. The analysis is mostly technical: the agency must determine whether the operator has provided all required documents, maps, and other information and whether they substantially comply with all reclamation standards. The plan cannot be denied on non-reclamation grounds—for example, if the proposed mining would be incompatible with surrounding uses or would create detrimental environmental effects. An agency wishing to deny a proposed mining application on those or similar grounds may do so only in conjunction with one or more of the discretionary entitlement applications associated with the reclamation plan, such as an application for a zoning change or a conditional use permit.

After concluding that the plan substantially meets all performance standards, the planning commission must certify that finding to the Director of the Department of Conservation. The commission then forwards the reclamation plan along with associated documentation, financial assurances, and CEQA documents to the Director for review. *Id.* § 2774(c). The plan is reviewed (by the Director or his or her designee), and written comments are prepared and provided to the lead agency within 30 days after receipt of the plan. *Id.* § 2774(d)(1).

The lead agency must then evaluate the Director's comments and respond within a reasonable period, addressing every major issue raised. *Id.* § 2774(d)(2). When there is disagreement, the agency must state in detail why the Director's comments or suggestions have not been accepted. *Id.* The Director must be given at least 30 days' notice of the time and place for the hearing on the reclamation plan or, if no hearing will be held, at least 30 days' notice of the agency's intent to approve the plan.

The SMARA statute does not make clear if a written response to the Director's comments must be approved after a second public hearing. Indeed, the statute only requires one public hearing, which must occur before departmental review to allow the lead agency to certify to the department that the plan meets all reclamation standards. *Id.* § 2774(a), (c). Therefore, it may be that a written response to the Director's comments and recommendations need not be approved in a public hearing, but may simply be prepared by staff.

However, a contrary position can be inferred from the model ordinance adopted by the SMGB, which requires a separate planning commission hearing after department review. Model SMARA Ordinance, § 6.0(i), (j). Given that the SMGB must ultimately certify the agency's mining ordinance, moreover, any ordinance the SMGB certifies may have to provide for two public hearings for reclamation plan review—one before and one after submission to the Department. The purpose of the second hearing would thus be to adopt responses to the Director's comments and finalize approval of the plan.

Whatever the case, after approving a reclamation plan, the lead agency must forward its response to the Director's comments within 30 days following that approval. Pub. Res. Code § 2774(d)(2). During that period, the Department retains all powers, duties, and authorities it holds under SMARA. *Id.*

Following approval of the plan, the lead agency must also record a "Notice of Reclamation Plan Approval" with the county recorder. *Id.* § 2772.7. The notice must read: "Mining operations conducted on the hereinafter described real property are subject to a reclamation plan approved by the [agency], a copy of which is on file with the [county recorder]." *Id.*

PRACTICE TIP

In some instances, lead agencies and other parties have asserted that information required to be included in reclamation plans, such as mining footprints, depths, and expected mining durations, define the extent of an operation's nonconforming uses under the Diminishing Asset Doctrine, discussed in chapter 9.

Operators of nonconforming operations should take great care during the reclamation plan approval process to prevent such arguments from arising after approval. To accomplish this, consider including language in the plan that all information provided is for purposes of reclamation and not an expression of the operator's intent as to the scope of its right to operate mining as a nonconforming use.

Appeals to the State Mining and Geology Board

When a city or county denies approval of a reclamation plan or fails to timely act, SMARA provides a mining operator a special avenue for challenging the agency's decision or inaction. In those situations, an operator may appeal to the SMGB, which, should it choose to grant the appeal, can approve the plan without having to remand the appealed matter back to the city or county.

Operators may appeal to the SMGB when their cities or counties deny plans that substantially meet SMARA standards, deny due process, or fail to timely act upon a reclamation plan application.

Appeal process. SMARA grants appellate jurisdiction to the SMGB to review appeals from any person claiming that the lead agency did not follow due process, or did not properly apply the "substantially meets" standard in denying approval of a plan. Pub. Res. Code § 2770(e)(1); Cal. Code Regs. tit. 14, § 3650(c)(1)–(3). An operator may also appeal to the SMGB if the lead agency fails to act within a reasonable time after receiving a reclamation plan application. Pub. Res. Code § 2770(e)(2); Cal. Code Regs. tit. 14, § 3650(c)(3). Notably, though, SMARA does not give the SMGB power to review a lead agency decision to approve a reclamation plan. Any person wishing to challenge the approval must therefore pursue a judicial remedy.

To perfect its appeal, the operator must file the appeal within 15 days of exhausting its rights to appeal with its city or county.

An operator filing a SMGB appeal must do so within 15 days of exhaustion of all remedies before the lead agency. Cal. Code Regs. tit. 14, § 3650. Upon receiving the operator's appeal, the SMGB chairman must then determine whether it falls within the SMGB's jurisdiction. *Id.* § 3651. The SMGB "*may* decline to hear an appeal if it determines that the appeal raises no substantial issues related to the lead agency's review...." *Id.* § 2770(f) (emphasis added). However, the SMARA regulations make clear that the SMGB Chairman "*shall* determine whether the appeal is within the jurisdiction of the...." Cal. Code Regs. tit. 14, § 3651. The regulations further provide that the SMGB may accept an appeal that (1) "raises any issues which can legally be addressed by [the SMGB]" or (2) "specifically relates to the lead agency's review of reclamation plans...." *Id.* § 3651(a), (b).

If the SMGB accepts the appeal, traditional administrative review procedures apply. An administrative record must be prepared that includes all documents the lead agency reviewed in denying or failing to act upon the reclamation plan. *Id.* § 3652(a). The SMGB then announces a notice of public hearing and holds the hearing on the appeal. *Id.* §§ 3655–3658. At the hearing, the appellant, lead agency, and members of the public may make statements in support of or in opposition. *Id.* § 3653(a). Department of Conservation staff may also be consulted in helping the SMGB resolve the appeal. *Id.* § 3653(a).

After closing the public hearing on the appeal, the scope of the determination is limited: the SMGB must only determine "whether the reclamation plan...substantially meet[s] the applicable requirements of [SMARA] and the lead agency surface mining ordinance...." Pub. Res. Code § 2770(g). Just as the lead agency's review of a reclamation plan is limited to determining compliance with reclamation standards and requirements, the SMGB may not approve or deny a plan based on a mining operation's perceived merits, impacts on surroundings, environmental impacts, or similar factors. The review is technical, to determine if the appealed reclamation plan substantially does or does not meet SMARA and its regulations.

If the SMGB determines that the appealed reclamation plan substantially meets SMARA's standards, it must approve the plan.

If the SMGB finds that the SMARA requirements are substantially met, the reclamation plan "shall be approved." *Id.* The SMGB thus has the final say once its appellate power is invoked. It cannot remand the appealed plan back to the lead agency, as an appellate court might do after overturning a trial court decision.

If it determines that the plan does not substantially meet all SMARA standards and requirements, the SMGB must return the reclamation plan to the operator with a notice of its deficiencies. *Id.* The operator then has 30 days to resubmit a corrected plan to the SMGB, unless the operator and SMGB agree on a different period. *Id.*

A prerequisite to judicial relief? Despite the specificity with which SMARA and its regulations prescribe the SMGB process, the Act does not make clear whether an appeal is required prior to filing a lawsuit challenging lead agency denial of a reclamation plan. The language is permissive, not mandatory, stating only that anyone entitled to appeal the denial of a plan "may" do so by filing an appeal with the SMGB. *Id.* § 2770(e). Regardless, California case law states that, "if an administrative remedy is provided by statute, relief must be sought from the administrative body and such remedy must be exhausted before judicial review of the administrative action is available." *Alta Loma Sch. Dist. v. San Bernardino County Comm. on Sch. Dist. Reorg.* (1981) 124 Cal. App. 3d 542, 554.

Moreover, because the full panoply of appellate procedures is part of the SMGB appeal process, and because an aggrieved party is entitled to full relief, it seems likely that the legislature intended that the appeal provisions be mandatory. *See Grant v. Super. Ct.* (1978) 80 Cal. App. 3d 606, 609. The composition of the SMGB, which includes geologists and other professionals with relevant technical backgrounds, bolsters this conclusion. By having the SMGB be the first body to hear the appeal—which often turns on interpretation of technical requirements for which a superior court judge probably would have little or no experience—the legislature likely intended to utilize the the SMGB's expertise in resolving reclamation plan appeals.

Although the issue is not settled, operators likely must exhaust their SMGB appeal rights as a prerequisite for bringing a lawsuit.

While an SMGB appeal may be requisite to the exhaustion of administrative remedies, the issue is not yet settled. The author therefore believes that anyone wishing to challenge the denial or the lead agency's failure to process the plan should simultaneously file an appeal with the SMGB and a petition for writ of administrative mandamus. *See generally* Code Civ. Proc. §§ 1094.5, 1094.6. This would allow the operator to exhaust all remedies, assuming a court later concluded that the SMGB appeal is mandatory. At the same time, this would allow an operator to defeat any statute of limitations defense the lead agency might assert for not filing a writ petition within 90 days of the lead agency's decision. *See id.* § 1094.6(b). If both remedies are pursued simultaneously, the court could always stay proceedings on the writ until the SMGB makes its final decision. As a practical matter, unlike a judicial appeal, the SMGB appeal can usually be resolved quickly. Given a crowded judicial docket, the appeal may even be final by the time the parties would be required to file their briefs in a court proceeding.

PRACTICE TIP

Operators should file both a SMGB appeal and a civil action within the applicable deadlines to ensure they do not waive any appeal rights.

Judicial Appeals

Anyone wishing to challenge the *approval* of a reclamation plan may only do so by filing a judicial appeal. As noted above, an operator may appeal to the SMGB but

Administrative mandamus is the proper procedure for challenging a city or county's approval of a reclamation plan.

only concerning a plan denial. A challenge to a plan approval must be accomplished by filing a petition for writ of administrative mandamus in superior court. *See generally* Code Civ. Proc. §§ 1094.5, 1094.6. Administrative mandamus is the proper and exclusive remedy for challenging local administrative decisions concerning land use applications. *State of California v. Super. Ct.* (1974) 12 Cal. 3d 237, 251.

The Supreme Court's El Dorado *decision confirms that the Director of the Department of Conservation may challenge city and county reclamation plan approvals.*

The Director of the Department of Conservation may likewise challenge lead agency approval of a reclamation plan. Until recently, whether the Director had that power was hotly contested. Noting that SMARA does not expressly give that power to the Director, a mining trade association intervened when the Director attempted to challenge a lead agency's reclamation plan approval in court. Ultimately the case made its way to the California Supreme Court, which ruled in favor of the Director. *People ex rel. Dep't of Conserv. v. El Dorado County* (2005) 36 Cal. 4th 971, 995. While lacking primarily responsibility for the Act's enforcement, the Court observed, the Director had several significant responsibilities in enforcing SMARA. *Id.* pages 990–91. Those responsibilities gave the Director a sufficient "beneficial interest" in the Act's enforcement to bring suit despite any express authorization to do so. *Id.* page 994.

Financial Assurances

Operators must post financial assurances to ensure that they, not the public, bear the expense of reclaiming mined land.

In addition to a reclamation plan, SMARA requires that a mining operator post a financial assurance. As noted above, this amendment was added to the Act in the early 1990s as part of a broader package of reforms designed to address lack of operator compliance and lead agency enforcement. Because a typical financial assurance can range from hundreds of thousands to millions of dollars, an operator now has a substantial financial stake in mine reclamation. Ignoring those responsibilities may cause an operator significant financial loss. At the same time, a mandatory financial assurance guarantees that an operator who fails to reclaim the operation does not leave a public agency—and ultimately taxpayers—with responsibility for reclamation. *See People ex rel. Dep't of Conserv. v. El Dorado County* (2005) 36 Cal. 4th 971, 991 ("A fundamental purpose of SMARA is that surface mine operators, rather than the taxpaying public, bear the expense of reclaiming land disturbed by surface mining.").

Form, Amount, and Duration of Financial Assurances

Financial assurances must take the form SMARA prescribes and remain in effect until reclamation is completed.

An operator must post a financial assurance only in forms provided for by SMARA or its regulations. The Act allows a financial assurance to take the form of a surety bond, an irrevocable letter of credit, or a trust fund. Pub. Res. Code § 2773.1(a)(1). The regulations also provide that a government operator may utilize a pledge of revenue and budget set-aside as a financial assurance. Cal. Code Regs. tit. 14, §§ 3803(b), 3806; *see also id.* §§ 3806.1 (pledges of revenue), 3806.2 (budget set-asides). Whatever form is used, the financial assurance must be payable both to the lead agency and the Department of Conservation. Pub. Res. Code § 2773.1(a)(4). It must remain in effect for the duration of the operation and, after completion of mining, until reclamation is completed. *Id.* § 2773.1(a)(2).

The amount of the financial assurance must be sufficient to complete reclamation per the operator's reclamation plan. *Id.* § 2770(d); Cal. Code Regs. tit. 14, § 3802(b). The amount must also include a reasonable estimate for administrative costs that the lead agency or Department of Conservation might incur in overseeing the project. However, the amount may not include the cost to complete mining of the site since that is not a reclamation-related expense. *Id.* § 3804(b).

The amount of the financial assurance must be sufficient to complete reclamation for the operator's reclamation plan.

In determining the specific costs of reclamation, a variety of factors are relevant. First, the costs must be based on the physical activities and materials necessary to implement the plan. *Id.* § 3804(a)(1). Second, the cost must account for any unit costs or third-party contracting costs that the lead agency may incur in overseeing the operator's compliance. *Id.* § 3804(a)(2). Third, the costs must account for the number of those unit costs. *Id.* § 3804(a)(3). Finally, as a precautionary measure against underestimated expenses, reclamation costs must include a contingency amount not to exceed 10 percent. *Id.* § 3804(a)(4).

Once the amount of financial assurance has been established, it must be adjusted annually. Pub. Res. Code § 2773.1(a)(3). The adjusted amount must account for new lands disturbed by the mining operation, inflation, and reclamation of land completed during the previous year. *Id.* To assist a lead agency in enforcing this requirement, an operator must submit a revised cost estimate each year to the lead agency. Cal. Code Regs. tit. 14, § 3804(c).

The amount of financial assurances must be adjusted annually to account for changes in the mining operations.

Financial Assurance Approval Process

The process for approving a financial assurance is the same as the one used for a reclamation plan. Thus, financial assurance approval occurs simultaneously with plan approval. A lead agency must ensure that the proposed financial assurance complies with the standards described above. As with a reclamation plan, the agency must also submit the financial assurance to the Director of the Department of Conservation for review and comment. Pub. Res. Code § 2774(c). Before approval of the financial assurance, the lead agency must—as with a reclamation plan—respond to all of the Director's comments within a reasonable period and specifically describe why the agency chose not to accept any of the comments. *Id.* § 2774(d)(2).

Financial assurances are processed simultaneously with reclamation plans.

The Department of Conservation reviews financial assurances just as it does with reclamation plans.

The "substantially meets" standard governs a lead agency's approval of a financial assurance. At a minimum, this requires that an agency find that the financial assurance is "sufficient to perform reclamation of lands remaining disturbed." *Id.* § 2770(d). If the evidence supports the finding, the agency *must* approve the financial assurance. As with reclamation plan approval, the agency's

Cities and counties must approve financial assurances if they substantially meet SMARA's requirements.

discretion is limited and cannot be used to deny approval for a reason unrelated to the sufficiency of the financial assurance.

Financial Assurance Appeals

Any operator who contends that a lead agency's denial of its proposed financial assurance was improper may, as with a denied reclamation plan, appeal to the SMGB. Pub. Res. Code § 2770(e). The standards and process for the appeal are virtually identical to those governing a reclamation plan. But while the process is the same, regulations governing the financial assurance appeal process appear in separate sections of the Code of Regulations. *See generally* Cal. Code Regs. tit. 14, §§ 3680–3690.

Operators may appeal denials of financial assurances to the SMGB in the same manner as for reclamation plans.

As with the reclamation plan appeal provisions, an aggrieved party may only appeal a financial assurance decision on the grounds that the lead agency relied on factors unrelated to SMARA or its regulations, denied due process, or failed to act within a reasonable period after receiving the financial assurance. Pub. Res. Code § 2770(e); Cal. Code Regs. tit. 14, § 3682(a)–(c). The SMGB's appellate jurisdiction is also limited. After accepting an appeal, the SMGB may only determine whether the financial assurance substantially meets SMARA and its regulations. Cal. Code Regs. tit. 14, § 3690. The SMGB must approve a financial assurance that meets all requirements. *Id.* As with a reclamation plan appeal, the SMGB may not remand the matter back to the lead agency for further action.

If it determines that a financial assurance substantially meets SMARA's standards, the SMGB must approve it.

Additional Operator Responsibilities under SMARA

An operator's responsibilities under SMARA are not met simply by securing approvals for a reclamation plan and financial assurance. Once those are approved, an operator has other significant and continuing duties, described below.

Duty to Amend a Reclamation Plan Before Deviation from Plan

Substantial deviations. Reclamation requires that a mine operator commit to a plan for mine closure that may not occur for many years or even decades. To engage in such planning with precision is difficult and, in some cases, impossible. Over the lifetime of the operation, assumptions that formed the basis for the original reclamation plan may change, particularly those related to the extent, quantity, and quality of the mineral deposits to be mined. At any point during the life of the operation, what the operator knows about the true yield of the mine's deposit may be based on estimates or models. Only after mining begins can an operator gain an accurate understanding of actual yield, both horizontally and vertically.

Surveying and exploration conducted after reclamation plan approval often identifies new areas and depths to which operators may mine.

Because the true extent of the deposit isn't always known, an operator cannot predict—and the reclamation plan may not identify—the full extent of surface area and depth an operator must mine. Based on surveying and exploration conducted after plan approval, an operator may later seek to expand the operation. If the operator does so, the changes may render the plan obsolete, making it difficult or impossible to reclaim the site in accordance with the plan.

Recognizing this possibility, SMARA does not mandate that a reclamation plan, once adopted, may never be changed. An operator may amend the plan when unexpected conditions arise. Pub. Res. Code § 2777. However, the Act requires that an operator anticipate the need for adjustments and obtain approval of an amended plan in advance of making changes. SMARA makes clear that "[s]ubstantial deviations from the original plan shall not be undertaken until [an] amendment has been filed with, and approved by, the lead agency." *Id.*

Operators seeking to substantially deviate from their reclamation plans must amend their plans before undertaking the deviations.

Despite the mandate to obtain approval of an amended reclamation plan, the Act did not define the term **substantial deviation**, making enforcement difficult. To compound the problem, another SMARA provision required that a plan approved prior to 1993, when substantially amended, had to conform to current reclamation standards. Cal. Code. Regs. tit. 14, § 3700(c).

Requiring that a plan conform to current standards—which are rigorous, comprehensive and costly to comply with—was a huge deterrent to operators seeking to amend their plans. Some would simply make the significant changes without obtaining lead agency approval. Consequently, plans would become outdated, with no measures for reclaiming the expanded mining areas or other changed conditions. The financial assurances, in turn, would become outdated, remaining at amounts well below those needed to fully reclaim the expanded operations.

Other operators, recognizing the need to reclaim their expanded mining areas, were more creative. Instead of amending the existing plan to meet all current standards, they prepared second reclamation plans that covered only the expanded areas. These operators essentially treated the expanded mining areas as separate operations, even though the purported separate operations occurred on the same property covered by the original plan. As a result, the costly and demanding modern standards were imposed only on the portions of the properties covered under the second plan.

In 2002, the SMGB adopted the One Reclamation Plan regulation, putting an end to this practice. This regulation generally prohibits a mine operator from having more than one reclamation plan for the same property. An operator proposing changes to a reclamation plan that constitute substantial deviation must amend the plan rather than prepare a separate plan for the expansion. *Id.* § 3502(d). Exceptions apply only to an operation that received approval of a second reclamation plan prior to adopting the One Reclamation Plan regulation or those with a single operation located in more than one jurisdiction. *Id.* § 3502(i).

In 2002, the SMGB adopted the One Reclamation Plan regulation, which prevents operators from obtaining approval of multiple reclamation plans for single operations.

Filling in a gap left by the earlier lack of definition, the One Reclamation Plan regulation defines "subtantial deviation" as "change[s] or expansion[s] to a surface mining operation that substantially affect[ ] the completion of the previously approved reclamation plan, or that change[ ] the end use of the approved plan to

Substantial deviations are changes that substantially affect completion of the reclamation plan, change the plan's end use, or substantially change the scope of its required reclamation.

the extent that the scope of the reclamation required for the surface mining operation is substantially changed." *Id.* § 3502(d). To determine if these conditions exist, a lead agency must consider a number of factors, including whether the changes in the operation (1) will deepen or substantially increase surface disturbance; (2) extend the termination date of the operation; (3) substantially affect the operation's post-mining use; (4) create any inconsistency with a previously-adopted environmental determination; or (5) create any other change the lead agency deems to be a substantial deviation. *Id.*

Application of the One Reclamation Plan regulation depends on the type of changes operators propose.

The extent to which an operator must amend a plan to comply with the One Reclamation Plan regulation depends on the type of change proposed. An operator might wish to extend the vertical or horizontal extent of an existing pit or excavation, or propose the creation of a new quarry pit or excavation entirely separate from the existing pit or excavation. The operator might further propose a change that does not require an extension or creation of a new excavation, but that would extend the lifetime of the operation or affect its post-mining use.

When an operator expands an existing excavation, it must amend the plan to make the entire area affected by the deviation comply with current SMARA standards.

An operator proposing a substantial deviation that involves expansion of an existing pit or excavation—by increasing the perimeter of the excavation area or by mining to a lower depth—must amend the existing plan to include current reclamation standards "for the entire area governed by the plan *that is impacted by the [substantial] deviation*." *Id.* § 3502(g) (emphasis added). Current standards do not apply to areas of the operation not affected by the deviation.

Unfortunately, the standards for determining whether and how an area of the operation is "impacted" for purposes of this requirement are unclear. The regulation does not define the term. While "impacted" would include the physical areas actually consumed by the laterally-expanded or deeper excavation, would it also include other areas of the overall pit or excavation? For example, if the lateral extension occurred only on the east side of a quarry pit, would the west side be "impacted"? Further, if a quarry pit is deepened, would areas above the point where the deepening occurred be "impacted" as well? Because the regulation does not answer these questions, operators and lead agencies may encounter difficulty when applying this aspect of the One Reclamation Plan.

Different standards apply if an operator proposes a substantial deviation that creates an entirely new quarry pit or excavation. For this type, the regulation offers two options. *Id.* § 3502(h). First, an operator may amend the existing plan to account for the expanded operation. In this case, current reclamation standards apply only to the area designated for use as a new quarry pit or excavation and any "processing facilities, roads, sumps, drainage systems or storage or processing areas" of the original reclamation plan area that will be used for the new pit or excavation. *Id.* § 3502(h)(1). Areas and facilities of the original plan area not utilized by the new pit or excavation may be accordingly reclaimed per the original plan. *Id.* With this option, however, the lead agency may require that the plan be amended so that

If an operator seeks to create a new excavation, it may amend its plan to apply current reclamation standards to the new excavation and all original plan areas the new excavation will affect.

an operator must immediately begin reclamation of any area the operator is no longer mining. *Id.*

Preparing a new reclamation plan for the new mining area is another option. *Id.* § 3502(h)(2). With this option, all facilities and areas existing under the original plan utilized by the new area must be reclaimed per current SMARA standards. However, the lead agency cannot require the operator to immediately begin reclaiming areas of the original plan that the operator is no longer mining.

An operator seeking to create a new excavation may also prepare a separate reclamation plan that covers the new area and all areas affected by it.

An operator whose substantial deviations do not involve either an extension of an existing pit or excavation or creation of a new pit or excavation must revise the entire plan to current standards. *Id.* § 3502(f). Thus, an operator proposing changes that would substantially affect the post-mining end use or extend the termination date of its operation must amend the existing reclamation plan to current standards before undertaking the changes. *See id.* § 3502(d)(2)(3).

Operators whose substantial deviations do not involve an expanded or new excavation must amend their entire plan to current reclamation standards.

Minor deviations. Under SMARA, deviations that do not meet the substantial deviation standard need only be submitted to the lead agency prior to making the changes; no lead agency approval is required. Pub. Res. Code § 2777. Further, current SMARA reclamation standards apply only to the amended portions of a reclamation plan, not to portions of the operation unaffected by the deviation. Cal. Code Regs. tit. 14, § 3502(f).

Operators need not obtain approval for non-substantial deviations but must submit plan amendments to their cities or counties.

Duty to Submit Annual Reports and Pay Annual Fees to the Department of Conservation

To assist the Office of Mine Reclamation's monitoring of compliance, an operator must annually submit a report to the Department of Conservation. In addition to basic information about an operator's property, operation, and type of material mined, the report must indicate:

Operators must submit annual reports to the Department of Conservation.

- That the lead agency has conducted its annual inspection of the operation
- That the operator has a financial assurance in place
- Total acreage of land that the operation disturbed during the previous year
- Total acreage reclaimed during the previous year
- Total unreclaimed acreage remaining at the end of the year
- Total production for each mineral commodity produced during the previous year

Pub. Res. Code § 2207(a)

An operator must also furnish this information to the lead agency, with the exception of proprietary mineral commodity production data, which may not be released to the public. *Id.* § 2207(b); *see also id.* § 2207(g) (prohibiting disclosure of operators' proprietary information, but providing for disclosure of statistics pertaining to total commodities produced per lead agency and statewide).

Operators must also pay an annual fee to the Department of Conservation.

In addition to providing the above information, an operator must pay an annual reporting fee to the Department of Conservation. *Id.* § 2207(d)(1). The fee is used to fund the Department's oversight of SMARA compliance as well as the Mineral Land Classification Program described in chapter 4. Fees for each mining operation are based on its size and type, total assessed value, acreage disturbed by mining activities, and acreage covered by the reclamation plan. *Id.* § 2207(d)(4).

Duty to Submit Interim Management Plan During Idle Periods

Mining is heavily dependent on the economy, and the need for minerals and construction materials generally follows economic trends. While it may soar during a sustained period of growth, mineral production may fall as the economy slows. During a prolonged downturn or recession, many operations—particularly smaller ones—may have to close temporarily due to lack of demand.

To avoid having to begin reclamation during temporary closures, operators must obtain approval of interim management plans.

Recognizing this possibility, SMARA provides for an idling period in which an operator may temporarily close without having to begin reclamation. To take advantage of this provision, an operator must submit and receive lead agency approval of an interim management plan. Pub. Res. Code § 2770(h)(1). Otherwise, an operation that remains idle for more than one year will be considered abandoned. *Id.* § 2770(h)(6). In that situation, an operator would be required to immediately commence reclamation or face forfeiture of the financial assurance. *Id.*

For purposes of SMARA's idle mine provision, an idle operation is one that curtails its activities by more than 90 percent of its previous maximum annual mineral production. *Id.* § 2727.1. The idling period must last for a year or more. *Id.* An operator must have an intent to resume operations at a future date, and cannot utilize the idling period to forestall its reclamation responsibilities. *Id.*

An operator must submit an interim management plan within 90 days of becoming idle.

An operator must submit the interim management plan within 90 days of becoming idle. *Id.* § 2770(h)(1). As with the approval of the original plan, a lead agency's power to review an interim management plan is strictly confined. Within 60 days of receiving an interim management plan, the agency must approve the plan "so long as the plan satisfies the requirements of [SMARA's interim management plan requirements]." *Id.* § 2770(h)(4). Unlike for a reclamation plan and financial assurance, however, no express provision of SMARA allows an operator denied the right to enter into an interim management plan the right to appeal to the SMGB. Nonetheless, an interim management plan is considered a reclamation plan amendment. *Id.* § 2770(h)(1). As an amendment, an unfavorable decision could likely be appealed under provisions that govern reclamation plan appeal. *See Id.* § 2770(e).

If a lead agency approves the interim management plan, the financial assurance must remain in effect during the entire idle period. Pub. Res. Code § 2770(h)(3). Once approved, the plan may then last for only five years. *Id.* § 2770(h)(2). After that time, a lead agency, finding that an operator has complied with the interim management plan, may renew it for an additional five years. *Id.* § 2770(h)(2)(A). Alternatively, the agency may require that an operator commence reclamation. *Id.* § 2770(h)(2)(B).

Duty to Complete Reclamation at the Conclusion of Mining

Once mining has ended, an operator's final duty under SMARA is to complete reclamation, and the financial assurance remains in place until the process is complete. Pub. Res. Code § 2773.1(a)(2). After the lead agency confirms that reclamation has been completed, the lead agency can release the financial assurance. To accomplish this, the SMARA statute provides that the lead agency notify the Director of the Department of Conservation "that reclamation has been completed in accordance with the approved reclamation plan." *Id.* § 2773.1(c).

To have their financial assurances released, operators must complete reclamation and their city or county must notify the Department of Conservation that reclamation is complete.

Despite significant objection by some operators, the SMGB in 2004 adopted a regulation specifying that an operator must also receive written concurrence from the Director of the Department of Conservation that reclamation has been completed. Cal. Code Regs. tit. 14, § 3805.5(d). Challenging its validity, an operator association subsequently filed suit, claiming that the regulation was inconsistent with SMARA's express requirement that the Director only *receive* lead agency notification of completion. The Association's position was ultimately rejected, resulting in affirmation of the SMGB's authority. *Mineral Ass'ns Coalition v. State Mining and Geol. Bd.* (2006) 138 Cal. App. 4th 574, 585.

in this chapter...

chapter 6

Enforcement of the Surface Mining and Reclamation Act

Cities and counties have a number of additional responsibilities under the Surface Mining and Reclamation Act (SMARA) besides ensuring that all mining operators have reclamation plans and financial assurances. Among other duties, they must adopt special mining ordinances, annually inspect mining operations, and ensure that operators annually review their financial assurances. Cities and counties have a variety of enforcement mechanisms to ensure compliance. And, if they fail to fulfill their duties or initiate appropriate enforcement, they can lose some or all of their enforcement powers to the State Mining and Geology Board (SMGB).

SMARA = Surface Mining and Reclamation Act

SMGB = State Mining and Geology Board

City and County Enforcement Responsibilities Under SMARA

Duty to Adopt a Surface Mining Ordinance

Before approving a reclamation plan, a city or county must adopt a surface mining ordinance. Pub. Res. Code § 2774(a). The ordinance must establish procedures for approval of a reclamation plan and financial assurance that in turn require at least one public hearing. *Id.* This ordinance must further be "in accordance with state policy." *Id.*; *see also generally id.* §§ 2755–2759. Consequently, at a minimum, a city or county must require reclamation of all mined lands within the agency's jurisdiction and the reclamation must meet the performance standards described in chapter 5. The ordinance must also require a financial assurance that meets all SMARA requirements and regulations. *Id.*

Cities and counties must adopt special surface mining ordinances that ensure reclamation and establish procedures for reviewing reclamation plans and financial assurances.

State policy, however, does not apply to "aspects of regulating surface mining operations which are solely of local concern...." *Id.* § 2757. Nothing in SMARA

The ordinance SMARA requires does not mandate how agencies must regulate local matters relating to mining.

limits an agency's power to regulate noise, dust, blasting vibration, hours of operation, pollution control, and like activities associated with mining. *See id.* (noting that matters of "local concern" include, but are not limited to, "hours of operation, noise, dust, fencing, and purely aesthetic considerations"). Whether to allow mining, and under what conditions mining may occur, are issues that SMARA does not address and need not be regulated by the ordinance.

The SMGB has adopted a Model Surface Mining Ordinance that many agencies have adopted as their surface mining ordinances.

A city or county surface mining ordinance has no prescribed form or content. As a practical matter, most ordinances closely track the language of SMARA. The SMGB has also adopted a Model Surface Mining Ordinance that an agency can adopt with few changes or additions, available from the Department of Conservation (www.consrv.ca.gov).

The SMGB must certify city and county surface mining ordinances.

Before enacting a surface mining ordinance, a city or county must submit the ordinance to the SMGB for certification. Pub. Res. Code § 2774.3. In making its decision, the SMGB must find that the ordinance "adequately meets, or imposes requirements more stringent than, the California surface mining and reclamation policies and procedures [described in chapter 5]." *Id.* Cities and counties may establish more stringent reclamation requirements so long as they do not prevent an operator from complying with SMARA. *See id.* § 2715(e).

If, after review, the SMGB finds a proposed mining ordinance deficient, the city or county must address the deficiencies upon the ordinance's return from the SMGB. *Id.* § 2774.5(a). If the SMGB then finds the revised ordinance deficient, the city or county is given one additional chance to comply. *Id.* § 2774.5(b). Should a lead agency fail to submit a compliant ordinance after its second submission, the SMGB can assume authority over the agency's reclamation plan approval until the lead agency submits a proper ordinance. *Id.* Whenever the SMGB approves a reclamation plan under its assumed authority, the city or county may not later modify the plan without the operator's consent. *Id.* § 2774.5(d).

Cities and counties need not adopt surface mining ordinances until they receive their first mining application.

Cities and counties need not adopt a surface mining ordinance if no active mining operations are within their jurisdiction. *Id.* § 2774(a). But once it receives an application for a mining permit, the agency must prepare and seek SMGB certification for a surface mining ordinance. *Id.*

Duty to Annually Inspect Surface Mining Operations

Cities and counties must annually inspect all mining operations within their jurisdictions.

After approving a reclamation plan, the city or county must monitor operator compliance with the plan and other SMARA requirements. In carrying out that responsibility, its most crucial duty is to conduct a regular annual inspection of all mining operations within its jurisdiction. Compliance with this duty is essential: without knowing the actual condition of each mining operation, a city or county cannot be certain that the required reclamation will occur. Pub. Res. Code § 2774(b).

If unaddressed for a long period, noncompliance with the reclamation plan can threaten an operator's ability to reclaim. If an operator has mined expanded areas not addressed in the original reclamation plan, for example, no measure would exist for reclaiming the expanded areas. More troubling would be the effect of that situation on an operator's financial assurance, which may remain well below what is needed to actually guarantee that the work is done. As a result, a city or county could ultimately be burdened by significant reclamation costs if an operator's financial assurance is insufficient to cover all costs and the operation is abandoned.

Without annual inspections, cities and counties risk later being stuck with the costs of reclaiming abandoned mining operations whose financial assurances are inadequate.

To ensure that inspections are carried out regularly, SMARA requires that cities and counties conduct an inspection once per calendar year for each operation within their jurisdiction. *Id.*; Cal. Code Regs. tit. 14, § 3504.5(a). The agencies must also conduct the inspections within six months after receiving an operator's annual inspection report. Pub. Res. Code § 2774(b); *see also* chapter 5, page 75 (discussing an operator's duty to submit an annual report). Cities and counties must also provide an operator with at least ten days' advance notice of an inspection and ensure that the inspectors conduct the inspection safely and in a manner that does not disrupt the mining operation. Cal. Code Regs. tit. 14, § 3504.5(a), (d).

✍ PRACTICE TIP

The annual inspection's importance cannot be overestimated. Cities and counties should not only ensure that they timely carry out their inspection responsibility, they should take great care in evaluating each inspection report to determine if any enforcement action is necessary.

The purpose of the inspection is "solely to determine whether the surface mining operation is in compliance with [SMARA]." Pub. Res. Code § 2774(b). In assessing compliance, the SMARA regulations make clear that the scope:

The purpose of the annual inspection is solely to determine whether the operation complies with SMARA.

> [M]ay include, but [is] not limited to the following: the operation's horizontal and vertical dimensions; volumes of materials stored on the site; slope angles of stockpiles, waste piles and quarry walls; potential geologic hazards; equipment and other facilities; samples of materials; photographic or other electronic images of the operation; any measurements or observations deemed necessary by the inspector or the lead agency to ensure the operation is in compliance with [the Act].
>
> Cal. Code Regs. tit. 14, § 3504.5(f)

SMARA regulations also impose certain restrictions on the inspectors who must "in the determination of the [city or county]" have "demonstrated competence in performing inspections of surface mining operations...." *Id.* § 3504.5(b). If an evaluation of geologic and engineering conditions is required, a state-licensed geologist or engineer must perform or supervise the inspection. *Id.* To avoid a conflict of interest, the inspector may not have worked for the operator for one year prior to the inspection. *Id.* § 3504.5(c). Nor may the inspector be an employee or have a financial interest in the operation. *Id.*

The inspector must have demonstrated competence in performing inspections of surface mining operations.

After an inspection, the inspector must summarize the findings on forms approved by the SMGB. *Id.* § 3504.5(g). Upon completion of the report, the city or county must provide the operator a copy of the report within 30 days. *Id.* The city or county must also notify the Director of the Department of Conservation within 30 days of the inspection, indicating whether an operator is in compliance with the reclamation plan and financial assurance. Pub. Res. Code § 2774(b).

The city or county must provide copies of the inspection report to the operator and the Department of Conservation.

Duty to Ensure That Operators Annually Revise Their Financial Assurances

Cities and counties must ensure that operators annually revise their financial assurances.

In addition to an annual inspection, cities and counties must also annually ensure that an operator revises the financial assurance "to account for new lands disturbed by surface mining operations, inflation, and the reclamation of lands accomplished in accordance with the approved reclamation plan." Pub. Res. Code § 2773.1(a)(3). To assist cities and counties in making these adjustments, an operator must annually submit a revised financial assurance estimate. Cal. Code Regs. tit. 14, § 3804(c). Cities and counties must use that revised estimate to determine the adequacy of the operator's financial assurance.

Duty to Forfeit Financial Assurances When an Operator Is Financially Incapable of Reclaiming or Abandons the Operation

When operators become financially incapable of reclaiming, SMARA requires that cities and counties forfeit the operators' financial assurances.

On occasion, a mine operator may abandon the operation or become financially incapable of completing reclamation. In those situations, SMARA requires that a city or county forfeit the operator's financial assurance and use its proceeds to reclaim the mine site. The SMARA statute does not specifically define the circumstances in which an operator, for purposes of forfeiture, has abandoned an operation or is financially incapable of completing reclamation. However, the regulations specify five circumstances in which forfeiture is appropriate:

SMARA regulations identify five circumstances in which operators are deemed to be financially incapable of reclaiming.

- The operator fails to actually post the financial assurance within 30 days after the lead agency approves the financial assurance, unless the operator has a financial assurance appeal pending. Cal. Code Regs. tit. 14, § 3811(a); *see* chapter 5, page 72 (discussing financial assurance appeals).
- The operator fails to provide the agency an annual revised cost estimate within 30 days of receiving notification that it must provide that estimate. *Id.* § 3811(b).
- The operator's financial assurance mechanism has lapsed and the operator has not replaced that mechanism with another suitable mechanism within 30 days, leaving the operator without sufficient remaining assurances to complete reclamation. *Id.* § 3811(c).
- The lead agency is unable to contact a mine operator or the operator's agent after 90 days of the mine becoming idle. *Id.* § 3811(d); *see* Pub. Res. Code § 2727.1 (defining circumstances under which a mine is considered idle).
- The operation remains idle for more than one year and the operator does not obtain approval of an interim management plan. Cal. Code Regs. tit. 14, § 3811(e).

Cities and counties must provide operators notice and a public hearing before revoking their financial assurances.

Before determining that an operator's financial assurance should be forfeited, the city or county must provide notice of the decision to the operator and conduct a public hearing. *Id.* §§ 3812, 3813. The city or county must follow its own procedures in providing notice. *Id.* § 3813. At the hearing, the agency must then determine whether any of the five circumstances described above continue to exist, making forfeiture necessary. *Id.* § 3815.

Once a decision has been made in favor of forfeiting the financial assurance, the city or county must again notify the operator, by personal service or certified mail, of its intention. Pub. Res. Code § 2773.1(b)(1). The city or county must give an operator 60 days to commence or cause the commencement of reclamation. *Id.* § 2773.1(b)(2). If an operator does not comply, the city or county must forfeit the financial assurance and use the proceeds to conduct reclamation in accordance with the operator's plan. *Id.* § 2773.1(b)(3), (4). The mineral and surface owners of the unreclaimed property must provide access to the agency to allow completion of the reclamation. *Id.* § 2773.2.

Following the public hearing, the city or county must notify the operator of its intent to forfeit the financial assurance and give the operator 60 days to commence reclamation.

Although forfeiture has a significant financial consequence for an operator, SMARA makes clear that an operator remains liable for any reclamation costs the city or county incurs beyond the amount of the forfeited assurance. *Id.* § 2773.1(b)(4). Whether the agency could actually recover the excess costs from an operator financially incapable of completing reclamation in the first place is a separate question. But the Act makes clear that an operator is liable for the full actual costs of reclamation, and the financial assurance does not establish a ceiling for that liability.

Operators remain liable for any reclamation costs that their forfeited financial assurances do not cover.

Duty to Provide the Department of Conservation Copies of Permits and Reclamation Plan Amendments

As of July 1991, SMARA required that a city or county submit to the Department of Conservation copies of permits, including conditions of approval, for every existing mining operation within its jurisdiction. Pub. Res. Code §2774(e). The Act also requires that a city or county send the Department by July 1 of each year copies of any amended permits, any permits for new operators, or a statement that no changes have occurred during the previous year. By July 1, a city or county must also provide copies of any reclamation plan amendments approved during the previous year. *Id.* Those responsible for SMARA compliance should take special note of these requirements since the Act expressly states that failure to comply "shall be cause for action under [the SMGB takeover provisions] of Section 2774.4." *Id.* *See* page 83 (discussing SMGB takeover of city and county enforcement authority).

Cities and counties must annually provide the Department of Conservation with amended permits or reclamation plans.

Duty to Adopt Procedures for Organized Recordkeeping and Monitoring

To assist in the responsibilities described above, the SMARA regulations mandate that each city and county "establish and maintain in-house measures and procedures to ensure organized recordkeeping and monitoring of surface mining reclamation under its jurisdiction." Cal. Code Regs. tit. 14, § 3504(a). The regulations, however, do not prescribe any specific format, methods, or procedures for the recordkeeping or monitoring.

Cities and counties must establish procedures to ensure organized recordkeeping and monitoring.

Duty to Take Enforcement Actions to Ensure Operator Compliance

Lead agencies must follow SMARA's enforcement procedures whenever they determine mining operations cannot be reclaimed in accordance with their reclamation plans.

A city or county's responsibility to administer SMARA does not simply mandate that reclamation plans and financial assurances be approved to fulfill the duties described above. The SMARA regulations go further, requiring that a city or county act affirmatively whenever an operator is not in compliance with the Act. In that situation, the lead agency *must* require that an operator amend the plan if mining operations "can no longer be reclaimed in accordance with [the] approved reclamation plan." Cal. Code Regs. tit. 14, § 3502(e). Public Resources Code §§ 2774.1 and 2774.2 provide specific enforcement procedures an agency must follow to carry out this duty. The relevant notice, hearing, and other procedural requirements are explained below.

✍ PRACTICE TIP

City and county staff and counsel should be aware that failure to commence enforcement proceedings against non-compliant operators is one of the most likely grounds for SMGB takeover of local SMARA powers.

The SMARA regulations also spell out adverse consequences for an agency's failure to take appropriate enforcement action. When a city or county finds that an operator is not in compliance, but does not act to correct the violation, the SMGB may assume one or more of its SMARA enforcement powers. *Id.* § 3805.5(e). SMGB takeover of lead agency authority, and its ramifications for cities and counties, are discussed below.

Administrative Enforcement Proceedings and Public Remedies Under SMARA

As part of the package of 1990s reforms discussed in chapter 5, SMARA contains detailed enforcement provisions authorizing both administrative and judicial remedies. However, these do not supersede any other remedies that may be available by law. Pub. Res. Code § 2774.1(g). The provisions below are not necessarily the exclusive remedies an agency has to enforce the Act.

Administrative Proceedings

SMARA provides its own comprehensive enforcement scheme in Public Resources Code §§ 2774.1 and 2774.2.

SMARA provides a comprehensive administrative scheme for determining whether an operator has violated the Act and for imposing fines to force compliance. Contained in Public Resources Code §§ 2774.1 and 2774.2, this scheme applies to a lead agency and also to the SMGB when it has assumed lead agency authority or has authority to bring an enforcement action under those sections.

The city, county, or Director must notify the operator of its violation by personal service or registered mail.

To begin enforcement proceedings, the city or county or—when initiated by the SMGB—the Director of the Department of Conservation must have determined that a surface mining operation is not in compliance with SMARA "based on [the] annual inspection" or "otherwise confirmed by an inspection" of the operation. Pub. Res. Code § 2774.1(a). In light of the latter, the city or county must conduct an inspection when information other than that contained in the annual inspection indicates that an operator is not in compliance with SMARA. The city or county must then notify the operator of the violation by personal service or registered mail. *Id.* After providing notice, it may issue an order, also by personal service or registered mail, that requires an operator correct its violations. *Id.* However, if the violation is that the operator does not have an approved reclamation plan or a financial assurance, the order may require that the operator cease all mining operations. *Id.*

If the operator does not cure its violation, the city, county, or Director may then issue an order to comply.

The order does not go into effect once it is issued. The lead agency or director must instead provide the operator a hearing and afford a reasonable time for an operator to comply. *Id.* § 2774.1(b). The order must specify the aspects of the operation that do not comply with the Act. *Id.*

The order does not take effect until the operator has been given a hearing and reasonable opportunity to comply.

The hearing may occur not earlier than 30 days after the order is issued. *Id.* If the city or county issues the order, the body that conducts the hearing is ordinarily the planning commission. This is because SMARA mandates that the order be appealable to a lead agency's "legislative body," discussed in greater detail below. *See id.* § 2774.2(a). If in the first instance a city council or board of supervisors hears the order, that is the highest body. An operator disagreeing with the disposition of the initial hearing would thus have no other avenue for appeal.

The public hearing concerning the appeal should be conducted by the planning commission.

If the city or county or Director determines at the hearing that the operator has in fact committed the violation or violations alleged, the operator must then comply with the order. However, because of ambiguities in the drafting of Public Resources Code § 2774.1, whether the operator must be given further time after the hearing to comply, or whether the order may take effect immediately, is not clear. According to the statute, the order must "specify a time for compliance which the lead agency or director determines is reasonable, taking into account the seriousness of the violation and any good faith efforts to comply with applicable requirements...." *Id.* § 2774.1(b).

PRACTICE TIP

To address an ambiguity in the drafting of Public Resources Code section 2774.1, cities, counties, and the Director should give an operator a reasonable period following conclusion of the public hearing to correct SMARA violations confirmed at the hearing.

The statute also states that the hearing shall be set "no sooner than 30 days after the date of the order." *Id.* It is conceivable that an agency could hold the hearing on or near the same date that it sets for the operator's compliance, as long as both dates are at least 30 days after issuance of the order. In that situation, the order could become effective on or near the same date as the hearing, leaving the operator with little or no time after the hearing to comply.

This could lead to a perverse result. An operator who believed in good faith that no provision of SMARA was being violated might not wish to assert that position fearing harmful consequences should the city, county, or Director disagree and the order were to go into effect immediately. Knowing that a wrong guess could lead to severe consequences, an operator would be faced with a Hobson's choice of abandoning the first position, that SMARA is not being violated, or raising a position that might be rejected without sufficient time to take whatever corrective action is required. This could subject an operator to substantial fines. To prevent such an anomaly, the operator should be given a reasonable time from the date of the order's issuance as well as a reasonable time for compliance after the hearing.

If an operator fails to comply with the order, it is subject to fines of up to $5,000 a day.

Once an order goes into effect, the consequences are significant. An operator is subject to fines of up to $5,000 per day, assessed from the original date of noncompliance. *Id.* § 2774.1(c). In determining the specific amount of penalty, however, the agency must consider "the nature, circumstances, extent, and gravity of the violation or violations, any prior history of violations, the degree of culpability, economic savings, if any, resulting from the violation, and any other matters justice may require." *Id.* § 2774.1(c). The statute provides that fines must be imposed by a separate order and must be served by personal service or separate mail. *Id.* The statute also provides that the order "may be imposed administratively." *Id.*

Operators may appeal a penalty order to the city council, board of supervisors, or SMGB.

Once issued, an administrative penalty order becomes effective immediately and must be paid within 30 days of issuance. *Id.* Nonetheless, the operator's duty to pay that amount is suspended if the operator appeals the penalty to the legislative body of the lead agency or, if the order was issued by the Director, to the SMGB. *Id.*; *id.* § 2774.2(a). On appeal, the legislative body (the city council or board of supervisors) or the SMGB may affirm, modify, or set aside the penalty order. *Id.* § 2774.2(c). It must further provide notice to the operator of its disposition of the appeal by personal service or certified mail. *Id.* § 2774.2(d).

Following denial of their penalty order appeals, operators may further appeal by filing a petition for administrative mandamus within 30 days.

Once the legislative body or SMGB has resolved the appeal, the administrative decision imposing the fines is final and may be appealed to a superior court. However, the statute provides that any action challenging an order must be filed within 30 days of the legislative body's or SMGB's decision. *Id.* § 2774.2(e). The statute also provides that the exclusive remedy for challenging such a decision is administrative mandamus. *Id.*; *see generally* Code Civ. Proc. §§ 1094.5, 1094.6. To perfect its appeal, moreover, the operator must pay the full administrative penalty, which is held in an interest-bearing account pending resolution of the appeal. Pub. Res. Code § 2774.2(d).

The SMGB may commence enforcement proceedings if, after notification, a city or county fails to commence enforcement proceedings on its own.

Whenever a lead agency exercises its enforcement powers—if the SMGB has not assumed that authority—the lead agency ordinarily has responsibility for initiating proceedings. Even when not exercising an agency's enforcement power, however, the SMGB may initiate proceedings under these sections. If it does, the SMGB must first notify the city or county in writing of an operator's violation, and the agency must not have taken appropriate enforcement actions within 15 days thereafter. *Id.* § 2774.1(f)(1). Alternatively, the SMGB may initiate enforcement proceedings when the Director determines that the violation amounts to an "imminent and substantial endangerment to the public health or safety, or to the environment." *Id.* § 2774.1(f)(2).

Judicial Remedies

In addition to the administrative enforcement proceedings described above, a city or county and the Director may seek judicial relief in certain instances. After determining that an operation is in violation of SMARA and that the violation "presents an imminent and substantial endangerment to the public health or environment," an action may be brought to abate the violation in any court of competent jurisdiction. Pub. Res. Code § 2774.1(d). If brought on behalf of the Director, the Attorney General must initiate the suit. *Id.*

Upon a complaint by the Director of the Department of Conservation or the SMGB, the Attorney General may also bring an action for administrative penalties for any violation of SMARA or its regulations. *Id.* § 2774.1(e). The Attorney General may initiate the action if he or she believes the violations have occurred. *Id.* The Attorney General may likewise bring an action to enjoin any violation of SMARA or its regulations. *Id.*

The Attorney General may also bring actions for administrative penalties or to enforce SMARA.

Removal of Operators from the AB 3098 List

An operator who does not have an approved reclamation plan and financial assurance faces an additional, and potentially very costly, consequence. Under the **AB 3098** provisions of SMARA, state and local agencies are prohibited from purchasing material or minerals from such an operator. Pub. Contr. Code §§ 10295.5(a), 20676. This prohibition also applies to contractors acting on behalf of, or pursuant to, a contract with a local agency. *Id.* § 20676(c). *See also id.* § 20676(b) (defining **local agency** to include "any county, city, whether general law or chartered, city and county, town, school district, municipal corporation, political subdivision, or any board, commission, or agency thereof....").

Each quarter, the Office of Mine Reclamation publishes a list—known as the **AB 3098 List**—of operators eligible to sell minerals and materials to state and local agencies. To avoid removal from this list, an operator must either have an approved reclamation plan and financial assurance or an appeal pending with the SMGB concerning its reclamation plan or financial assurance. *Id.* §§ 10295.5(a), 20676(a).

Operators without approved reclamation plans or financial assurances may not sell to state or local agencies under SMARA's AB 3098 provisions.

State Mining and Geology Board Takeover of Lead Agency Authority

As noted in chapter 5, the early experience with SMARA's enforcement saw the Act go mostly unenforced. Many agencies failed to require that an operator have a reclamation plan. Even after approving a plan, some cities or counties devoted little effort to monitoring compliance. These problems led to significant revisions to the Act, with provisions giving the SMGB power to take over some or all of a city or county's SMARA administrative powers. The SMGB may take over any or all or the city or county's powers when it finds that the agency has:

As a safeguard for ensuring that cities and counties enforce SMARA, the Act allows the SMGB to take over SMARA enforcement powers from cities and counties whose enforcement is deficient.

- Approved reclamation plans or financial assurances that do not comply with SMARA or its regulations
- Failed to annually inspect operations within its jurisdictions
- Intentionally misrepresented its inspections

- Failed to forfeit financial assurances and to use the proceeds of those assurances to reclaim mining operations
- Failed to take appropriate enforcement actions
- Intentionally misrepresented the result of required inspections
- Failed to submit information to the Department of Conservation as required

Pub. Res. Code § 2774.4(a)

Before taking over any city or county power, the SMGB must give cities and counties 45 days to correct the deficiencies in their enforcement and hold a public hearing.

Before assuming any of these authorities, the SMGB must notify the city or county of its intention, giving the agency 45 days to correct its deficient enforcement. *Id.* § 2774.4(c). If the city or county does not correct the deficiencies within that period, the SMGB may hold a public hearing after providing mailed notice to every operator within the jurisdiction and notice by newspaper publication. *Id.* When deciding to assume administrative power, the SMGB must adopt written findings addressing the basis for its decision as well as any significant issues raised by the city or county, all operators, and any person who presented testimony or evidence at the hearing. *Id.* § 2774(e). Judicial review of the SMGB's decision may thereafter be sought by a petition for administrative mandamus so long as the petition is filed within 30 days of its decision. *Id.* § 2774(f).

After assuming any or all of a city's or county's powers, the SMGB must exercise those powers for at least three years. *Id.* § 2774.4(b). To restore the powers to the lead agency, the SMGB must hold a hearing and determine that the city or county has corrected deficiencies that led the SMGB to assume enforcement. *Id.*

The SMGB cannot assume any city's or county's permitting authority.

Regardless of the powers assumed, the SMGB may not remove a lead agency's permitting authority. *Id.* § 2774.4(a). Even when all its SMARA powers are assumed, a lead agency still retains power over permitting and zoning matters relating to mining operations.

Part Three

Local Permitting

in this chapter...

chapter 7

City and County Land Use Regulation

Part Two addressed the Surface Mining and Reclamation Act, California's major statutory scheme for regulating mining. Now the book turns its attention to city and county land use regulation. Because surface mining is such an intensive activity, cities and counties regulate it to ensure compatibility with its surroundings, and that impacts on neighboring land uses are minimized. As a result, it is subject to extensive regulation—through general plans, zoning ordinances, and conditional use permits.

Three Levels of Regulation

Mining is governed by three levels of local land use regulation: the general plan, zoning ordinance, and conditional use permit.

Like all other land uses, three levels of local regulation govern surface mining. The first, the general plan, is the overall blueprint for where, how, and under what standards a particular land use may occur. The second, the zoning ordinance, implements the broad mandates of the general plan by classifying land uses into separate districts, each with minimum requirements. The third and most specific, the conditional use permit, enables agencies to tailor the regulation of individual land uses to the peculiarities and needs of specific locations. Because the policies, standards, and requirements reflected in these three levels of regulation vary among local agencies, it is impossible to offer a uniform description of how agencies regulate mining. Nonetheless, city and county regulations share a common thread, and this chapter discusses typical regulations that one might find at the local level.

First Level—The General Plan

The first and broadest level of local land use regulation is the general plan. Before an agency's zoning code may authorize mining, and before an agency may issue a permit to mine, its general plan must identify the proposed site as a permissible location for mining, and the proposed operation must comply with all general plan standards. The zoning ordinance and permit authorizing the mining must then be consistent with the general plan.

The general plan is the constitution and charter guiding long-term growth within a city or county.

What is a general plan? Each city and county must have a general plan that provides an overall blueprint for management of land uses within its jurisdiction. Gov't Code § 65300. Serving as a master plan for development, the general plan has been called both the charter and constitution for local land use decisions. *City of Santa Ana v. City of Garden Grove* (1979) 100 Cal. App. 3d 521, 532; *Lesher Commun., Inc. v. City of Walnut Creek* (1990) 52 Cal. 3d 531, 540. Under California's Planning and Zoning Law, the plan must be "comprehensive" and "long term." Gov't Code § 65300. It must also include a "statement of development policies" as well as text and diagrams setting forth "objectives, principles, [and] standards" for development. *Id.* § 65302.

General plans are broken down into discrete elements, including the land use and conservation elements, which are the ones most likely to affect mining.

The style, format, and structure of a general plan vary among jurisdictions. Nonetheless, for every jurisdiction, state law requires that a general plan be separated into discrete elements that correspond to specific aspects of growth and development. These include elements for land use, circulation, housing, conservation, open space, noise, and safety. *Id.* §§ 65301, 65302. Each element must also meet minimum statutory requirements. *See Camp v. Bd. of Supervisors* (1981) 123 Cal. App. 3d 334, 352 n.10. A general plan's failure to meet those requirements can result in the invalidation of any ordinance or project approved in accordance with that plan. *Resource Def. Fund v. County of Santa Cruz* (1982) 133 Cal. App. 3d 800, 806; *Neighbor Action Group v. County of Calaveras* (1984) 156 Cal. App. 3d 1176, 1187.

This chapter discusses the two elements of a general plan related to the regulation of mining: land use and conservation. It also addresses the **consistency doctrine,** which requires that all zoning and permit decisions be in agreement with the general plan. *See Curtin's California Land Use* (28th ed. 2008, Solano Press) by Cecily Talbert, which explains a general plan's overall requirements, the required elements, the process for adoption or amendment, and methods for and consequences of a general plan challenge.

General plan requirements for mining uses

The land use element identifies land uses that are permissible within city or county limits and specifies the geographic distributions of those uses.

Land use element. Although every element is important, the land use element garners the greatest attention. That should not be surprising, since its purpose is to determine which land uses can go where. Under the State Planning and Zoning Law, the land use element must identify all uses of land that are permissible within the jurisdiction, and specify their geographic distribution. Gov't Code § 65302(a). This is accomplished with descriptive text and maps.

Before an agency can grant a permit, the land use element must designate the proposed area as a proper location for mining. General plan land use categories are typically broad, so this usually means that the proposed area will fall within a category permitting industrial uses. In some jurisdictions, a land use element will have a special designation for mineral resources. In others, mining may even be allowed in areas classified for open space or agricultural use.

Before mining can occur, the general plan must identify the mine site as one appropriate for mining.

Whatever general plan designation allows mining, the proposed operation must also comply with all policies the land use element establishes. The element does not simply determine where a land use may occur, but prescribes policies governing that use. These policies may deal with any number of subjects, focusing on broad overall goals rather than details. Usually they require conditions or mitigation measures that address land use incompatibilities between neighboring uses.

Conservation element. Surface mining often implicates the policies of the general plan's conservation element. As explained in chapter 4, a lead agency must adopt mineral resource management policies for any area the state classifies or designates as containing significant deposits. *See* Pub. Res. Code §§ 2762–2763. Designed to protect mineral resources from incompatible development and encroachment, these policies are often included in the conservation element. Cal. Code Regs. tit. 14, § 3676(c). Yet, while its focus is often on the impacts neighboring development may have on potential extraction, the element sometimes imposes restrictions on mining operations, such as noise buffers and setbacks. Adherence to conservation element policies, like compliance with land use element policies, is therefore often a prerequisite for a proposed mining project.

Mining must comply with all policies the conservation element establishes, such as those relating to land use compatibility.

Consistency doctrine. The general plan has primacy in all local land use matters. To ensure this effect, the Planning and Zoning Law requires that every subordinate land use action be consistent with the general plan. Gov't Code § 65860. Under the consistency doctrine, all ordinances, enactments, and project approvals must conform to the general plan's goals, policies, and standards. For a mining project, this doctrine requires that the next two levels of regulation—zoning and the conditional use permit—be in accordance with all relevant portions of the general plan. That is, all pertinent policies of the land use and conservation elements, along with any other applicable element, must be incorporated into the zoning and permit conditions for a mining project.

Under the consistency doctrine, all subordinate land use actions must be consistent with the general plan.

To be consistent with the general plan, an enactment or action need not be an exact match. *Sequoyah Hills Homeowners Ass'n v. City of Oakland* (1993) 23 Cal. App. 4th 704, 717. It must only be "in agreement or harmony" with the plan. *Greenebaum v. City of Los Angeles* (1984) 153 Cal. App. 3d 391, 406. Overall, "[a]n action, program, or project is consistent with the general plan if, considering all its aspects, it will further the objectives and policies of the general plan and not obstruct their attainment." *Corona-Norco Un. Sch. Dist. v. City of Corona* (1993) 17 Cal. App. 4th 985, 994 (*quoting* Governor's Office of Planning and Research, *General Plan Guidelines,* 212 (1990)).

Land use actions must be in harmony with the general plan to satisfy the consistency doctrine.

Second Level— The Zoning Ordinance

As the first level of regulation, the general plan is broad, setting out overall land use policies that leave out much of the detail about implementing those policies. That detail is filled in beginning with the next level of regulation, the zoning ordinance. If the general plan is the "constitution" or "charter" of land use controls, the zoning ordinance can be considered the statutes or bylaws.

Zoning is the regulation of the use of land, size of buildings, and developable areas of parcels.

What is zoning? Zoning is the regulation of land use, the size of buildings, and the developable areas of legal parcels. *O'Loane v. O'Rourke* (1965) 231 Cal. App. 2d 774, 780. Embodied in municipal ordinances, zoning regulations prescribe specific standards for development and buildings within a municipality. Zoning also classifies the various permitted land uses by type, separating them into discrete districts. Through zoning, a land use may be carried out only if it takes place in a district that allows that use. Specific remedies and penalties are provided to ensure compliance.

Now the common and primary feature of local land use regulation, zoning was not always so pervasive. The first comprehensive zoning ordinance was not adopted until 1916 by New York City. 1 *Anderson's Am. Law of Zoning* § 3:7 (4th ed.). In California, many local agencies enacted zoning ordinances several years later, some in the 1930s and 1940s. Even then, the ordinances may have had only limited applications; many properties were not placed in zoning districts until the 1950s or 1960s.

Zoning is the principal land use control in California.

For at least the last half century, though, zoning has been the principal land use control in California. This is no doubt due to its considerable advantage over other controls, such as the law of nuisance and restrictive covenants. By their nature, these land use controls are focused on individual situations involving a small number of parties. Because they further depend on voluntary private conduct—the filing of a lawsuit or the transfer of real property—their use is too sporadic and inefficient to allow for a comprehensive system of control.

With the pervasiveness of zoning today, land use can be controlled universally and uniformly. Zoning ensures—in theory, if not always in practice—that incompatible land uses will be separate from one another, prohibiting the creation of nuisances. A sensitive use, such as a residence and school, is separated from an intensive use, like a factory or an industrial operation. In short, the idea behind zoning is to keep out the "pig in the parlor," as one landmark case has observed. *Vill. of Euclid v. Ambler Realty Co.* (1926) 272 U.S. 365, 388.

Cities and counties often allow mining in only a few zoning districts.

Zoning of mining operations. Because it is so intensive, most zoning ordinances confine mining to a single or only a few districts that permit industrial uses. Some urbanized counties allow mining in an agricultural district, while other jurisdictions create a special **mineral extractive** or **overlay** zone. As discussed in chapter 4, these zones are an appropriate way to protect areas containing mineral resources the state has classified as significant. *See* Cal. Code Regs. tit. 14 § 3676(c)(3)(A) (providing that a lead agency may protect a significant mineral resource area by placing it in a special overlay zone).

Whatever type of district allows mining, mining occurs in areas removed or sufficiently buffered from sensitive land uses. Because of noise, dust, blast vibration, and other impacts associated with extraction and processing, mining is not suitable for residential, commercial, and other populated zoning districts.

Zoning schemes usually view mining as unsuitable for residential, commercial, and similarly populated zoning districts.

Conditional Use Permit

The final level of local land use regulation, the **conditional use permit** (CUP), is the most specific. While filling in the details of a general plan's policies, a zoning ordinance cannot account for the peculiarities of an individual location and its surroundings. Some uses, like mining, conflict with a surrounding land use no matter where they are located. The conditional use permit addresses those situations, allowing an agency to tailor land use approvals to the needs of particular sites.

The conditional use permit allows cities and counties to tailor their regulation of mining to the needs and surroundings of particular locations.

CUP = Conditional use permit

Permitted use versus conditional use. Each zoning district specifies certain allowable uses, in one of two ways. It may specify a **permitted use** that a person may automatically engage in once the minimum prescribed standards are met. If the applicant satisfies every requirement, a local agency may neither deny the use nor subject it to the public hearing process. Rather, by complying with every applicable standard, a person may engage in the permitted use as of right.

Permitted uses are ones that zoning districts allow so long as the particular uses meet prescribed standards.

Alternatively, the zoning district may permit certain uses on a conditional basis. When an applicant shows that possible land use incompatibilities can be mitigated, a **conditional use**, generally considered inappropriate for a particular district, is allowed. Because of the potential for a land use conflict, a conditional use is not permitted as of right, but instead requires a site-specific approval. Moreover, a conditional use is not only subject to a public hearing process, but may be approved or denied at the discretion of the local agency.

Conditional uses are ones that require discretionary city or county approvals.

A CUP usually imposes several restrictive conditions tailored to the specific location where the use will occur. For instance, it may impose hour restrictions, limit noise to certain levels, or require buildings and operations to be set back from property lines. It may impose a time limit for the use, after which a new permit must be issued, and can also impose a wide range of other conditions.

SMARA's permit requirement. In almost every zoning ordinance, mining is only allowed as a conditional use. Because it can adversely affect the surroundings, few zoning ordinances deem surface mining appropriate as merely a permitted use, for which there is no hearing process or discretionary approval. In fact, an ordinance permitting mining as of right likely conflicts with state law.

Zoning ordinances that allow mining as a permitted use likely conflict with state law.

The State Surface Mining and Reclamation Act, discussed in chapters 3 through 6, requires that every mining operation obtain a permit from the lead agency. Pub. Res. Code § 2770(a). SMARA defines **permit** as "any authorization from, or approval

Because SMARA requires a permit for every mining operation, it likely requires that cities and counties allow mining as only a conditional use.

by, a lead agency, the absence of which would preclude surface mining operations." *Id.* § 2732.5. Although not elaborating further, the definition seems to suggest that an agency must require something more than a determination that the use meets the criteria for a permitted use.

Had its intention been otherwise, the legislature would not have expressly required a permit through SMARA. In fact, at the time the Act was adopted, land uses were already allowed either as-of-right or upon issuance of a CUP, and a permit requirement that encompassed both would have been superfluous. By requiring a permit, the legislature was probably trying to preclude an agency from allowing mining as a permitted use. To comply with SMARA, therefore, any zoning ordinance that authorizes mining must do so only upon the issuance of a permit.

Nothing in the Act mandates that conditions be imposed as part of the permit, and its definition does not address the topic. Other provisions also make clear that SMARA does not restrict lead agency zoning and permitting powers. *See id.* § 2715(e) (confirming that SMARA does not preempt a lead agency's zoning powers); *id.* § 2757 (noting that state policy for mined lands does not "include aspects of regulating surface mining operations which are solely of local concern... such as... dust, noise, fencing, and purely aesthetic considerations.); *Id.* § 2774.4(a) (preventing the State Mining and Geology Board from assuming "permitting authority" from a lead agency). Thus, while it must require that a mining operation obtain a permit, an agency may freely address those matters through the exercise of its zoning powers.

Once an operator incurs substantial expenses in reliance on a CUP, it acquires a vested right that cities and counties cannot take away absent compelling circumstances.

Rights conferred by a conditional use permit. Once issued, a CUP confers important protections. A permittee who has diligently commenced operations or has incurred substantial expenses in reliance on the CUP acquires a vested right to continue operating. *Korean American Legal Fdn. v. City of Los Angeles* (1994) 23 Cal. App. 4th 376, 393 n.5. *But see Trans-Oceanic Oil Corp. v. City of Santa Barbara* (1948) 85 Cal. App. 2d 776, 783 ("If the permittee does nothing beyond obtaining the permit... the proper authorities may revoke it."). Once the permittee's rights vest, due process restrictions apply, and the power of a city or county to revoke the permit is limited. *Trans-Oceanic*, 85 Cal. App. 2d 783. An agency may revoke a permit only if the permittee fails to comply with its terms or there is a "compelling public necessity warranting the revocation," such as when the permittee carries on the business in a manner that creates a nuisance. *O'Hagen v. Bd. of Zon. Adjust.* (1971) 19 Cal. App. 3d 151, 158–159.

Zoning ordinances customarily provide for revocation or modification of CUPs under certain circumstances.

Many zoning ordinances expressly identify circumstances that allow revocation of a CUP. Typically, this includes when the permittee violates one or more of its conditions, creates a nuisance through its activities, or fails to disclose material facts at the time of the application. Changed circumstances since approval of the permit may also warrant revocation, but many ordinances provide for modification as a less drastic alternative.

Common Local Mining Regulations

Every city and county has a general plan, zoning ordinance, and permitting scheme, but the policies and standards for each may differ among agencies. Yet, agencies typically impose the following regulations, which are by no means exclusive, on surface mining.

Noise

Surface mining can be noisy. As dozers and loaders extract and transport material for processing, accelerating engines and beeping from backup alarms, along with clangs as rocks are loaded onto conveyors and processing equipment, can often be heard from nearby properties. The processing equipment itself emits steady operating noises as it crushes and screens excavated material.

Because mining is noisy, zoning ordinances usually establish specific noise limits.

Because mining operations are noisy, zoning ordinances or CUPs usually specify limits. Some agencies measure noise at the property line of the mine, while others use the property line of the nearest receptor. Specific allowable limits and the method of measurement vary. Some agencies average the noise over a period of one or two hours, and others use a daily average or specify the maximum single noise in a day. However it is done, one should expect some form of noise restriction when dealing with the permitting of mining operations.

Trucking Operations

The most common complaints about surface mining relate to trucks hauling the mined material along public roads. The number of trucks entering and leaving a mine site, especially during a peak morning commute, can be substantial as job sites typically require early deliveries. The persistent rumble of trucks may upset residents having to drive behind them. Residents may also complain about deterioration of the roads and the dust and debris of heavy traffic.

While a number of cities and counties regulate trucking aspects of mining, they often do so without complying with the California Vehicle Code.

Given these concerns, many zoning ordinances and conditional use permits attempt to restrict trucking activities associated with mining—through trip limits, road impact mitigation fees, haul route designations, and dust and debris restrictions. However, the California Vehicle Code expressly preempts all local regulation of these matters, unless specifically authorized. Veh. Code § 21. Although not every heavy truck restriction is barred, the Vehicle Code does prohibit some regulations that agencies may try to impose.

Truck trip limits. Perhaps the most effective way an agency can deal with the impacts of hauling is to specify the number of trips allowed. While agencies may wish to eliminate or minimize truck-related noise and safety impacts, however, trip limitations can cause a mine operator considerable difficulty, especially for a very large project or during busy seasons when a substantial number of trips are necessary. Whatever the merits or hardships, truck limit restrictions are probably not

Truck trip limits likely conflict with Vehicle Code section 21, which preempts local traffic regulation of matters the Vehicle Code covers unless expressly authorized.

lawful. The Vehicle Code comprehensively regulates the use of the state's roads and highways and, as part of its comprehensive scheme, the state expressly preempts local traffic regulation, providing through Vehicle Code section 21 that:

> Except as otherwise expressly provided, the provisions of this code are applicable and uniform throughout the State and in all counties and municipalities therein, and no local authority shall enact or enforce any ordinance on the matters covered by this code unless expressly authorized herein.

As noted by the courts, when the Vehicle Code covers a subject, its preemption by local regulation is presumed unless the legislature declares otherwise. *Barajas v. City of Anaheim* (1993) 15 Cal. App. 4th 1808, 1818. To avoid preemption, a local agency either must not address any matter the Vehicle Code covers or, if it does, the code must expressly authorize the matter.

Because the Vehicle Code regulates truck traffic in a number of ways, any local regulation of truck traffic likely would constitute a subject covered by the Vehicle Code.

Since heavy truck traffic is already regulated, limiting the number of trips is likely a covered subject. The Vehicle Code establishes the maximum weight that can be hauled. *See generally* Veh. Code §§ 35550–35796. Cities and counties are permitted to close certain streets and highways to heavy truck traffic. *See id.* §§ 35701, 35706. Because truck traffic is subject to these rules, *any* local regulation of heavy vehicle traffic would likely encompass a matter the Vehicle Code covers. As a result, section 21 preemption would probably apply, and an agency would be confined to adopting or enforcing an ordinance regulating heavy truck traffic only in ways that the Code "expressly authorize(s)." *Id.* § 21.

Although not tested in the courts, nothing in the code allows a municipality to regulate the number of trucks permitted for a particular land use. Indeed, the code allows a local agency to impose very few regulations. In addition to the statutes mentioned, an agency may restrict the hours trucks may operate on certain streets. *Id.* § 9400.7(a). But no provision allows an agency to limit the *number* of truck trips associated with a particular use.

Because of the lack of express Vehicle Code authorization, any municipal truck trip limits are probably unlawful.

Because of the lack of express authorization, any ordinance empowering a city or county to impose truck trip limits is likely unlawful. The same would be true of a CUP restriction on the number of trips. Section 21 applies not only to the adoption of an ordinance covered by the code, but also to its "enforce[ment]." By imposing a CUP restriction on truck traffic, an agency would necessarily be enforcing the zoning ordinance authorizing requirement of a permit. An operator could agree to abide by trip limitations, as is done on occasion to maintain comity with the local agency and good public relations. Absent voluntary compliance, however, truck trip limitations are likely unenforceable as in conflict with Vehicle Code section 21.

Road impact fees. Another regulation cities and counties impose to offset the impacts of hauling mined material is a road mitigation fee. Agencies often charge a fee for repair and upkeep of roads damaged by the high volume of heavy truck traffic, but these fees are under the same legal cloud as are truck trip limits.

Under the preemption mandate of section 21, a road mitigation fee is clearly a matter covered by the Vehicle Code. The code already requires that heavy commercial vehicles pay weight fees. *See* Veh. Code § 9400. Rather than authorizing adoption of additional fees, moreover, the code expressly prohibits them. Section 9400.8 states:

Vehicle Code section 21 precludes cities and counties from imposing road impact fees for the truck traffic from mining operations.

> Notwithstanding any other provision of law, if the voters approve Senate Constitutional Amendment 1 of the 1989–90 Regular Session, no local agency may impose a tax, permit fee, or other charge for the privilege of using its streets or highways, other than a permit fee for extra legal loads, after December 31, 1990, unless the local agency had imposed the fee prior to June 1, 1989.

With the approval of Proposition 111 in the June 1990 election, the condition that Senate Constitutional Amendment 1 be adopted was fulfilled. Thus, the command of section 9400.8 is clear: No local agency may impose any fee on mining operations "for the privilege of using [the agency's] roads or highways."

Because fees designed to recover road maintenance costs or mitigate other impacts associated with trucking would burden a mining operator for the "privilege" of hauling material along a public road, section 9400.8 precludes these fees, no matter how beneficial or necessary. *See County San. Dist. No. 2 of Los Angeles County v. County of Kern* (2005) 127 Cal. App. 4th 1544, 1618–1622 (invalidating an impact fee charged to recover roadway damage costs from the hauling of biosolids). An operator could, of course, volunteer payment of a fee or not challenge its validity to maintain a good relationship with a governing agency. From a legal standpoint, however, the Vehicle Code expressly preempts such a fee.

Haul routes. A local agency will often regulate the route a truck must travel from the mine site to a major public road or highway, known as a **haul route**. Unlike truck trip limits and road impact fees, regulating a haul route is generally permitted. However, haul route regulations must be adopted in the manner the Vehicle Code prescribes. Section 35701 permits haul route authorization, providing that "[a]ny city, or county for a residence district, may, by ordinance, prohibit the use of a street by any commercial vehicle or by any vehicle exceeding a maximum gross weight limit. . . ."

The Vehicle Code does not preclude cities and counties from designating the haul routes that mining trucks may travel, but cities and counties must follow the correct procedures for establishing them.

Under the code, a county may also "by ordinance reduce the permissible weight of vehicles and loads upon unimproved county highways or upon county bridges." *Id.* § 35706. By ordinance, a county may further reduce the maximum vehicle weight on improved highways if the highways will deteriorate due to the volume of heavy truck traffic. *Id.* § 35707.

Each regulation must be adopted "by ordinance." Therefore, an agency must restrict haul routes legislatively, not simply by imposing a regulation through a CUP condition, as many agencies try to do. This distinction may seem meaningless, but compliance with the Vehicle Code requires that an agency adopt a regulation only as "expressly authorized." *Id.* § 21.

Cities and counties must adopt haul routes by ordinance.

Cities and counties cannot use their Vehicle Code authorization to prohibit truck travel on the only road or roads for which mining operations can reach major thoroughfares and highways.

An additional consideration is that the designated route must afford a mining operator viable access to major roads and thoroughfares. Despite its ability to limit truck traffic on certain roads, an agency cannot use the Vehicle Code to prohibit truck travel on the only road or roads an operation may have to reach a major road or highway. *See Neary v. Town of Los Altos Hills* (1959) 172 Cal. App. 2d 722, 726. In rejecting one agency's attempt to do this, a court explained:

> We realize that it is the desire of the town to maintain itself for residence purposes only, without any commercial business within its limits, and that the driving of heavy trucks through a residence is highly undesirable. On the other hand, no municipality can make itself an island through which ordinary traffic (which includes heavy trucks) may not pass.
>
> *Id.* 728

While they must allow a mining operation access to public roads, cities and counties need not always allow an operator to have the most direct access.

Yet, while it cannot restrict all access to public roads, an agency need not allow a mining operator to have the most direct access to a main thoroughfare and highway. Heavy truck traffic may be prohibited on certain roads, so long as a mine operator retains viable access to major public roads and highways. *See Skyline Materials, Inc. v. City of Belmont* (1961) 198 Cal. App. 2d 449, 457. Even when trucks are required to travel additional distances or more circuitous routes, designated routes that impose some hardship on an operator are lawful. *Id.* 457–458.

Cities and counties often require that trucks wash their wheels, travel on paved or gravel-lined areas, or cross rumble grates to control trackout of dirt and debris on public roads.

Loading, washing, and debris control. Because trucks leaving a mine site are loaded with rocks, gravel, sand, and other earthen materials, they may track dirt and debris onto public roads. To address **trackout**, an agency may require that trucks pass through a wheel washer or rumble grate; or that surfaces separating truck loading areas and public roads be paved or lined with gravel, allowing caked-on mud to fall off truck tires. Some agencies even have operators run street sweepers along designated public roads.

The Vehicle Code prohibits cities and counties from requiring that trucks cover or tarp their loads.

Requiring a cover or tarp is another preferred regulation. Technically, the Vehicle Code requires that aggregate truck loads be covered. Veh. Code § 23114(e)(1). However, this requirement does not apply to truckers if they cover loads with six inches of open space at the top of a trailer container, called **freeboard.** *Id.* § 23114(e)(4). In addition, the peak of a load cannot extend above the upper edge of the container. *Id.* Given this exception, a local regulation that requires covering would be expressly preempted by the Vehicle Code.

Hours of Operation

To protect neighboring land uses, cities and counties typically limit the hours that mining operations may operate.

Agencies often regulate the hours of a mining operation when a sensitive use is nearby, effectively protecting adjacent property from early-morning and late-evening noise. Hour restrictions may also indirectly protect residents living along a public road from the noise of heavy traffic, because trucks cannot enter or exit a mine site when the mine is not operating. The permitted hours vary depending upon an agency's preferences and local conditions.

Hour restrictions can adversely affect a mine operator who has to begin early and end late to meet construction demands. At the same time, though, nearby residents may prefer more controls. Some of the tension between operators' and residents' concerns can be addressed by allowing a certain number of days each year for trucking to extend beyond its normal schedule. Many agencies provide for such extended operations, usually with advance notice.

Dust

Dust from mining is another area of concern. To control migration onto neighboring properties, many agencies require that an operator pave or place gravel on top of the onsite roads used by their dozers, loaders, and other heavy equipment. Agencies also usually require that an operator periodically water dirt roads, stockpiles, and other working areas. These measures often overlap with conditions imposed by air pollution control districts. These permits are discussed in chapter 10, Air Quality.

Cities and counties require that mining operators implement a number of measures to control dust.

Blasting

Some surface mines use explosives to break up hard-rock deposits. For these operations, a local agency often imposes regulations to protect neighboring residences from the vibration associated with blasting. An operation may be required to install an onsite or offsite seismograph to ensure that individual blasts not exceed acceptable limits, or the blasting may have to occur at limited times on weekdays, when the largest number of people are not home.

Mine operators that engage in blasting must usually monitor the vibration from their blasting and blast only at certain times.

Lighting

Where mining occurs after dark, the light required for the operation can be distracting to nearby residents. Many agencies require that an operator install lighting that is not visible beyond the property line or to neighboring properties.

Many agencies require that lighting used for mining operations not be visible to neighboring residences.

Screening and Fencing

For mining located near a residential area, many agencies require that plants, equipment, and other facilities be screened from view. For aesthetic reasons an agency may require that the operator not exceed maximum stockpile heights, and that some or all of the property have landscaped berms, trees, or vegetation. The mined property may also have to be fenced off for reasons of safety and the restriction of public access.

Erosion and Sedimentation

To prevent a mining operation from causing erosion of adjacent properties, an agency sometimes requires the construction of settling ponds, installation of dikes or ditches, planting of vegetation, or similar measures. Sediment and erosion control plans may also be required. Often, these requirements incorporate by reference

Cities and counties impose a number of requirements on mining operations to protect water quality.

waste discharge requirements and storm water pollution prevention plans imposed by a regional state water quality board. These authorizations are discussed in chapter 11, Water Quality.

Local Permitting Process—an Overview

Readers may be familiar with the permitting process for mining applications, or at least with general land use approvals. For others, the process—with its formalities, acronyms, and specialized planning and environmental terminology—may be difficult to understand. For these readers, the discussion below provides a brief, nontechnical overview. The first section explains the specific approvals, often called **entitlements**, that are required to permit a new mining operation. The second describes the beginning-to-end process in which cities and counties decide to approve or deny a project. The third section places the local permitting process in context, identifying the many other authorizations an operator must obtain from various agencies before beginning to mine.

Land Use Entitlements That an Applicant May Request

A mining application generally involves up to three potential requests: a general plan amendment, zoning change, and conditional use permit.

Generally, an applicant seeking to mine will request as many as three land use entitlements from a city or county. These include a general plan amendment, a zoning change, and a CUP. Which entitlements are requested depends on the general plan designation and zoning classification governing the proposed site at the time of the application.

When the general plan does not allow mining at the location of the proposed mining operation, the applicant must request a general plan amendment.

A general plan amendment is necessary if the land use element does not designate the proposed site as one where mining is permissible. Since it is the "charter" and "constitution" for long-term development within a city or county, the general plan must authorize mining within its overall plan for growth. Moreover, because of the consistency doctrine, any zoning classification or conditional use permit that allowed mining would be void absent authorization by the general plan.

If the zoning of the proposed mine site does not allow mining, the applicant must request either a zone change or a zoning text amendment.

Just as the general plan must designate the proposed site as an appropriate location for mining, so too must the site's zoning district. When mining is not allowed, the zoning must be changed to permit a mineral-extractive use. This request may take two forms. An applicant can request that the site's zoning be changed to allow mining. This is known as a **zone change** or **rezone**. Alternatively, an applicant can request a zoning **text amendment** that would not change the site's zoning district, but would amend the district's governing regulations to make mining a permissible use.

Applicants must always request conditional use permits as a part of their mining applications.

Even if the general plan and zoning district for a proposed site already authorize mining, all applicants must still—because of SMARA's "permit" requirement, as discussed above—obtain a CUP. At a minimum, therefore, a mining application will always involve a CUP request.

Review of Mining Applications

To begin the process for entitling a surface mining project, the operator must file an application with the governing city or county for each land use entitlement required for the proposed site. At the same time, the operator must submit a reclamation plan application and proposed financial assurance to comply with all reclamation obligations, as described in chapter 5. Although land use and reclamation entitlements regulate different aspects of surface mining—the land use entitlements address the matters described in this chapter, and the reclamation entitlements address the post-mining conversion of a mine site to its subsequent use—cities and counties process the entitlements simultaneously throughout the overall application process.

Filing a mining application is usually immediately followed—and often preceded—by a meeting between an applicant, the consultants, attorneys, and the city or county planning staff responsible for review. The initial task is to ensure that the application contains all of the plans, drawings, studies, and other information staff needs to evaluate the project. The process may last a number of months, with staff regularly working with an applicant to obtain everything required to process the application.

After filing an application, city or county staff must first determine that the application is complete.

Once staff deems that the application is complete, the environmental review process begins. (Mandated by the California Environmental Quality Act or CEQA, the requirements are discussed more fully in chapter 8.) At this stage, staff must determine if any aspect of the proposed operation would create a significant and adverse environmental impact requiring the preparation of an environmental impact report (EIR), a document that analyzes the scope of these impacts and identifies the measures, if any, necessary for mitigation. To accomplish this task, staff prepares a document know as an **initial study**, and circulates the document to departments within the city and county and to other potentially-affected governmental agencies. Following receipt of their comments, staff then determines the level of environmental review necessary. As explained in chapter 8, this involves whether to prepare an EIR, a full environmental report, or a negative declaration, a less involved document appropriate for a project that would not create any significant and adverse environmental impact. Because of the intensive and industrial nature of mining, a negative declaration for mining is unusual; most agencies choose an EIR for such a project.

After deeming the application complete, staff must circulate an initial study to determine the level of CEQA review required.

CEQA = California Environmental Quality Act
EIR = Environmental impact report

The CEQA document is then prepared and circulated for public review. If staff determines that an EIR is required, which is usually the case, this step can require a significant amount of time—usually at least a year and, in some cases, two or more years. Generally, an environmental consulting firm is retained to prepare the document, which is then circulated for mandatory public comment. Following public comment, the application is ready for public hearing, which in almost all California cities and counties is held before the planning commission.

Following preparation of the CEQA document, the application is ready for a public hearing before the city or county planning commission.

While city and county practices vary, one of the planning commission's first tasks usually deals with reclamation and does not involve approving or denying the requested land use entitlements. As discussed in chapter 5, SMARA requires that

Figure 7-1
Local Permitting Process

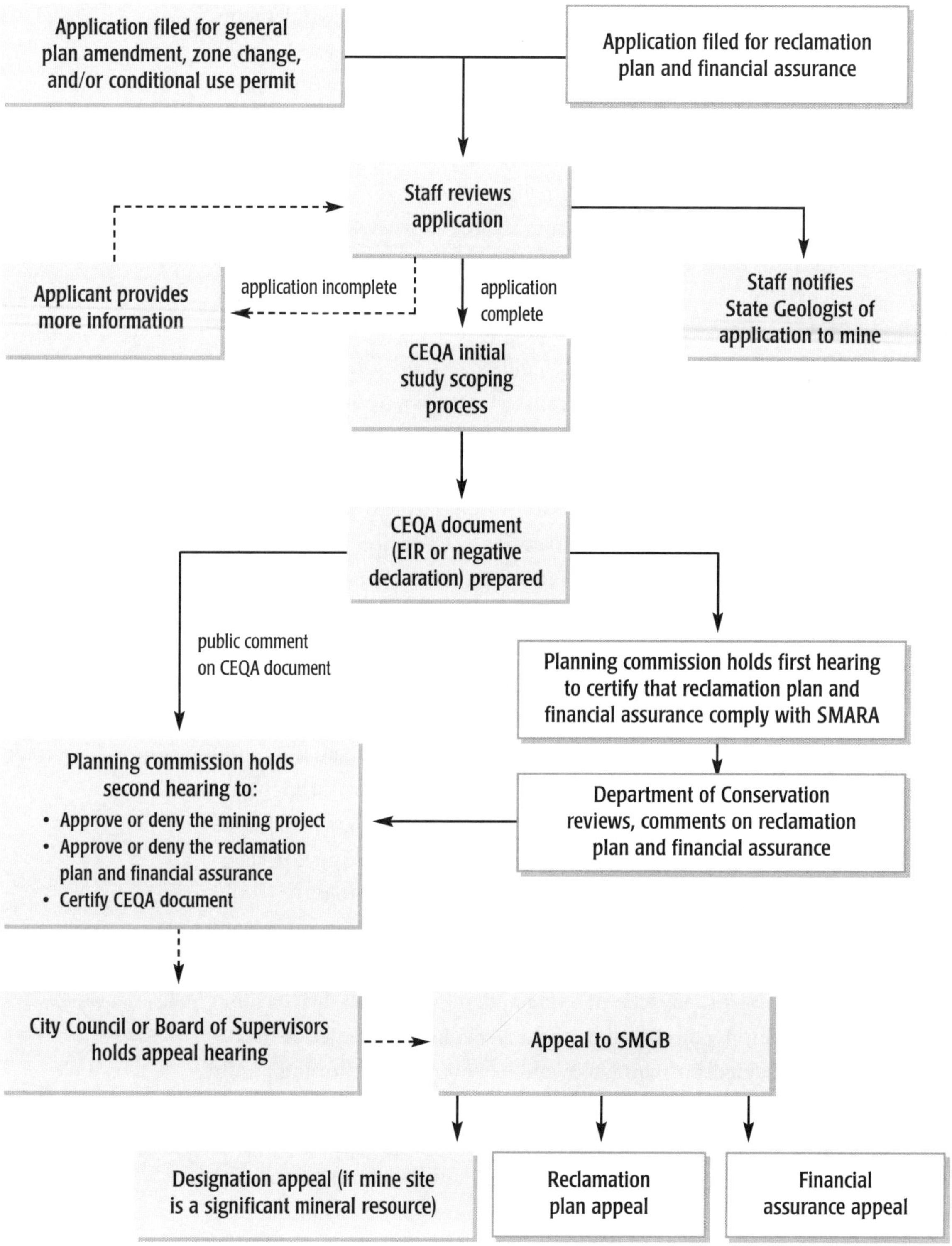

the planning commission certify to the SMGB that the reclamation plan and financial assurances meet all of the Act's requirements. Often, this is done by the planning commission at an initial public hearing. The commission then forwards the plan and the assurance to the Department of Conservation for review. Once the Department evaluates and comments on the plan and assurance, SMARA requires that the planning commission respond, either by accepting the Department's recommendations or by providing a statement of its reason for rejecting them. Cities and counties often do this at a second public hearing.

Part of the planning commission's task in reviewing an application is to determine in a public hearing that the reclamation plan and financial assurances comply with SMARA.

At this second public hearing the planning commission may decide whether to approve the overall mining project—by approving or denying the general plan amendment, zoning change, and/or the conditional use permit. At this hearing, the planning commission must also certify the environmental document prepared for the project.

At a second public hearing, the planning commission decides whether to approve the mining project and certifies the CEQA document.

Whatever the outcome, the decision is almost always appealed to the city council or board of supervisors. Indeed, to preserve the right to later appeal the project decision to a court, any parties aggrieved by the planning commission's decision must appeal. When they do, the council or board then itself considers some or all aspects of the proposed project, depending on the specific issues appealed. The council or board considers the appeal in a separate public hearing, following which it affirms or overturns the planning commission decision, thus deciding in effect whether to approve or deny the application.

Once the council or board has acted, dissatisfied parties may challenge the decision in court. They may seek to overturn the decision through a special procedure know as administrative mandamus, or claim that it was not made in compliance with CEQA. An appellant may also allege inverse condemnation on the grounds that the city or county has effectively taken the property without just compensation. In certain situations described in chapter 4, and also explained below, parties may appeal to the SMGB, claiming that the decision was not supported by substantial evidence.

If any party disagrees with the council's or board's decision, it must then appeal to the superior court or to the State Mining and Geology Board.

Other Applications

Besides city and county review of the land use entitlements and SMARA requirements, an operator must also simultaneously pursue a number of authorizations from other governmental agencies. These include air quality permits (chapter 10), NPDES permits or waste discharge requirements (chapter 11), Section 404 permits and streambed alteration agreements (chapter 12), endangered species "take" permits or habitat conservation plans (chapter 13), and hazardous materials business plans (chapter 14). Each agency has its own process for issuing the authorizations, as explained in Part 4, Environmental Regulation. While retaining *principal* authority, cities and counties are not the *only* agencies with authority over the proposed mining project, but are usually the "lead agencies" for conducting environmental review under CEQA. That is, the other agencies will use the environmental document a city or county prepares as a basis for evaluating the environmental effect of the permits they consider.

Mining operations require a number of governmental authorizations from agencies other than cities or counties.

NPDES = National Pollutant Discharge Elimination System

Special Procedures Governing Mining Applications

With two exceptions, the procedures governing review of mining applications are the same as those for land use applications generally.

With two exceptions, the procedures governing mining applications are the same as for land uses. As with land uses, an agency must provide notice to the public of hearings concerning the mining application and hold a public hearing before making a decision. *See Horn v. County of Ventura* (1979) 24 Cal. 3d 605, 612–613; *Scott v. City of Indian Wells*, 6 Cal. 3d 541, 549 (1972). An agency must also support its decision by preparing written findings explaining the basis for the decision. *See Topanga Ass'n for a Scenic Cmty. v. County of Los Angeles* (1974) 11 Cal. 3d 506, 515. Findings must further be supported by substantial evidence in the record of documents, staff reports, testimony, and other evidence presented to the agency. *See Citizens Ass'n for Sensible Dev. of Bishop Area v. County of Inyo* (1985) 172 Cal. App. 3d 151, 173; *Placer Ranch Partners v. County of Placer* (2001) 91 Cal. App. 4th 1336, 1341–1342. These notice, hearing, and finding requirements are discussed in depth in *Curtin's California Land Use and Planning Law* (Solano Press, 28th ed. 2008). In addition to these general requirements, SMARA imposes two special requirements for mining applications that are described below.

Notification to the State Geologist of Surface Mining Applications

Cities and counties must notify the State Geologist within 30 days of receiving an application to mine.

Cities and counties must notify the State Geologist of any application to mine land in their jurisdiction within 30 days of receiving the application. Pub. Res. Code § 2774(e). When an agency fails to comply with this notification requirement, SMARA provides that the SMGB can assume its SMARA administration powers. *Id.* § 2774.4; *see* chapter 6, page 87 (describing SMGB assumption of lead agency authority).

Permit Appeals to the State Mining and Geology Board

Following a city or county decision concerning a permit to mine, an aggrieved party may appeal to the SMGB, which must then determine if substantial evidence supported the city's or county's decision.

The decision of a city or county concerning an application to mine on lands designated by the SMGB to contain a significant mineral resource can be appealed to the SMGB. Pub. Res. Code § 2775(a); *see also* chapter 4 (discussing city and county responsibilities for protecting significant mineral resources). Unlike a reclamation plan appeal to the SMGB, addressed in chapter 5, the permit appeal right applies both to an applicant challenging a permit denial and an individual challenging its approval. Pub. Res. Code § 2775(a). However, the SMGB's authority over such appeals is limited. After following standard hearing procedures, the SMGB may only determine whether substantial evidence supported the agency's denial or approval. *Id.* § 2775(c). If it finds the evidence lacking, the SMGB may then only direct that the city or county reconsider its decision. *Id.* It cannot reverse the decision, either by affirming a permit approval or by overturning a denial. These **designation appeals** are discussed further in chapter 4.

Special Issues Created by Mining Applications

Application of Local Regulations to Mining on Federal Land

The type of land use regulations described in this chapter apply to federal as well as private land. The U.S. Supreme Court confirmed the power of local agencies to regulate mining on federal land in *California Coastal Commission v. Granite Rock Company* (1987) 480 U.S. 572. Although that case involved application of State Coastal Commission regulations, the Court held that Congress did not preempt state—and by implication—local regulation of mining through federal mining laws. *Id.* 589.

The United States Supreme Court's Granite Rock *decision confirms that cities and counties may regulate mining on federal land within their territories.*

Granite Rock involved an application to mine property on federal forest land, authorized by the Federal Mining Act of 1872 and other laws. Absent from those laws, the Court observed, was any evidence that Congress sought to restrict states from regulating the environmental impacts mining may have. Instead, Congress had distinguished between regulations for the environment and those for land use planning. *Id.* 588. The latter—concerning *whether* a person may mine—might be preempted, but the former—concerning only *how* mining may occur—were not. *Granite Rock* held that a local agency may only impose what the Court characterized as environmental regulations on federal property and may not impose "land use planning" regulations.

Granite Rock *held that cities and counties may impose environmental regulations on federal land but not land use planning regulations.*

Despite asserting this distinction, *Granite Rock* noted that "[t]he line between environmental regulation and land use planning will not always be bright; for example, one may hypothesize a state environmental regulation so severe that a particular land use would become commercially impracticable." *Id.* 587. Nonetheless, the Court clarified that "the core activity described by each phrase is fundamentally different. Land use planning in essence chooses particular uses for the land; environmental regulation, at its core, does not mandate particular uses of the land but requires only that, however the land is used, damage to the environment is kept within prescribed limits." *Id.*

Following *Granite Rock's* environmental-land use planning dichotomy, the Eighth Circuit Court of Appeals in *South Dakota Mining Association, Inc. v. Lawrence County* (8th Cir. 1998) 155 F. 3d 1005, held invalid a local ordinance prohibiting new surface metal mining operations in national forest land. In so holding, the court observed that:

> The Lawrence County ordinance is a per se ban on all new or amended permits for surface metal mining within the area. Because the record shows that surface metal mining is the only practical way any of the plaintiffs can actually mine the valuable mineral deposits located on federal land in the area, the ordinance's effect is a de facto ban on mining in the area. Thus, unlike *Granite Rock*, we are not faced with a local permit law that sets out reasonable environmental regulations governing mining activities on federal land.
>
> *Id.* 1011

As South Dakota Mining *clarifies, the land use planning regulations preempted under* Granite Rock *are those that, in effect, amount to a de facto ban on mining.*

As *South Dakota Mining* clarifies, the land use planning regulations preempted under *Granite Rock* are those that, in effect, amount to a "de facto ban" on mining. Regulations that allow mining, but that govern the conditions under which it occurs, are not preempted. Thus, as long as local regulation of mining on federal property does not prohibit mining, the regulation is equally enforceable as if it were imposed on private property.

Mining on Land Under a Williamson Act Contract

Mining operations are sometimes proposed on land under a Williamson Act contract, a special designation given to land devoted to agriculture, recreation, and open space. Gov't Code § 51201(d). In exchange for preferential tax treatment, a property owner agrees through a Williamson Act contract to carry on certain uses that are compatible for preserving agricultural land. *Id.* §§ 51238.1(a), 51240, 51243(a). In return, a local agency ensures that the agricultural resources under contract will not be prematurely developed, since a Williamson Act contract must remain in effect for at least ten years and automatically renews each year unless the property owner elects not to renew. *Id.* § 51244.

A number of issues arise when a mining operation is proposed on land covered by the Williamson Act. Foremost is whether the mining can even occur as long as the contract is in effect. Under the Act, only agricultural and compatible uses are allowed. *Id.* § 51243. Because it is obviously not agricultural, mining can accordingly be permitted only when deemed a "compatible use."

Whether mining is a compatible use depends on the contract date. As a result of legislative changes that became effective in 1994, the Williamson Act now expressly recognizes mining as a compatible use for certain post-1994 contracted lands, but imposes onerous conditions in doing so. To be compatible, mining on post-1994 contract land must be referenced within the "four corners" of the governing contract. *Id.* § 51238.3. At a minimum, the mining may not significantly impair the underlying contractual commitment to preserve prime agricultural or open space land. *Id.* § 51238.2. To comply with this requirement, the mined property must usually be reclaimed to agricultural land after being mined.* *Id.*; *see also* chapter 5 (discussing reclamation of mined lands). Significantly, this means that the land must meet SMARA's demanding standards for agricultural land reclamation, which require that:

- Land to be reclaimed as prime agricultural land be restored to a "productive capability... equivalent to or exceed[ing], for two consecutive crop years, that of the premining condition...."

* The land may also be reclaimed for open space use. However, the land must be restored only to the specific "open space" uses specified in the Williamson Act. These include lands within the scenic highway corridor, wildlife habitat areas, saltponds, managed wetland areas, and submerged areas. Gov't Code § 51201(o).

- Land to be reclaimed as non-prime agricultural land be restored "so as to be capable of sustaining economically viable production of crops commonly grown in the surrounding areas."

Cal. Code Regs. tit. 14, §§ 3707(c), 3708

The Williamson Act makes clear that there is no exception to these standards. Gov't Code § 51238.2.

In sum, for mining to occur on post-1994 contract land, an operator must show that the property will be reclaimed to sufficient soil capacity for a post-mining agricultural use. An operator would thus most likely be required to backfill all disturbed areas, including quarry pits and working areas, with soil capable of meeting the SMARA agricultural land reclamation standards. If the mining operations will last for an extended period and involve significant surface disturbance, that may be difficult to accomplish.

For mining to occur on post-1994 contract land, an operator must show that the property will be reclaimed to sufficient soil capacity for a post-mining agricultural use.

For a pre-1994 Williamson Act contract, a compatible use includes only one that was deemed compatible at the time the contract was submitted or signed. *Id.* § 51238.3(a). Whether mining was considered compatible before 1994 is unlikely, since Williamson Act policies were intended to preserve a "maximum amount of the limited supply of agricultural land"; to discourage "premature and unnecessary conversion of agricultural land to urban uses"; and to preserve open space, scenic highway corridors, and wildlife habitat." *Id.* § 51220(a), (c)–(e). Not only is mining not even mentioned, additional support for that conclusion stems from the legislature's adoption of SMARA in 1975, ten years after adoption of the Williamson Act. SMARA creates its own scheme for conserving mineral resources, demonstrating an implicit legislative treatment of mining and agriculture as distinct. *See generally* chapter 4 (discussing SMARA's protection of significant mineral resources).

If mining is not considered a compatible use—either because it cannot meet the post-1994 standards or would occur on pre-1994 contracted lands—another option would be to cancel the contract. Ordinarily, a Williamson Act contract automatically renews unless the property owner elects not to renew it. *Id.* § 51244. Because the minimum term is ten years, a property owner usually must wait at least nine years before implementing a non-Williamson Act use. Upon cancellation, however, the Williamson Act contract is immediately terminated.

Nonetheless, the standard for cancellation is high. An agency may only terminate a contract if doing so would be consistent with the purposes of the Act or would be in the public interest. *Id.* § 51282. The California Supreme Court has held that these standards require "extraordinary situations." *Sierra Club v. City of Hayward* (1981) 28 Cal. 3d 840, 864. Terminating a contract simply to allow a mining operation to commence would likely be difficult.

in this chapter...

chapter 8

Surface Mining and the California Environmental Quality Act

by Jennifer T. Buckman and James D. McNairy

Since surface mining uses land intensively and often requires discretionary entitlements from governmental agencies, mining operations are subject to the California Environmental Quality Act (CEQA). Enacted in 1970 with the goal of protecting the environment, CEQA is one of the most demanding environmental laws passed by any state.

Mining operations use land intensively and often require many permits that may make them subject to CEQA.

Surface mining, unlike a less-intensive land use, can substantially affect natural resources and the environment. A mining operation may require many acres of land, continue for decades, and occur in stages. Moreover, mining uses large amounts of water and generates noise and dust on an ongoing basis, characteristics that trigger the need for detailed environmental review.

CEQA = California Environmental Quality Act

Generally speaking, the more intensive the land use, the greater the scrutiny that CEQA requires. This chapter discusses the type of review required for mining projects and recurring issues associated with the environmental analysis of reclamation plans to restore land once mining operations are complete.

Brief Overview of CEQA

The CEQA statutes, guidelines, and case law are voluminous, and a full review of the Act is beyond the scope of this book. Instead, this chapter focuses on special issues related to surface mining—in particular, how to determine if CEQA applies to a proposed mining activity and the general requirements the law imposes. For a thorough discussion of the Act, readers should consult Ronald E. Bass *et al., CEQA Deskbook* (Solano Press), which provides an excellent introduction to its requirements and procedures, and Tina A. Thomas *et al., Guide to the California Environmental Quality Act* (Solano Press), widely regarded as the comprehensive summary of the law.

Overview of the CEQA Process

CEQA establishes a three-step process for a lead agency to determine the appropriate level of environmental review for a particular project under consideration. For a mining project, the lead agency is the city or county that has principal responsibility for approving the project. *See* CEQA Guidelines § 15367.

The first step is to determine if the activity is subject to CEQA review. Although rare, some projects do not trigger CEQA. For example, an operation that was fully entitled and permitted before CEQA was enacted may not require additional discretionary action by a public agency and, hence, may not be subject to review.

If the activity constitutes a project subject to CEQA, the second step is to determine whether any of CEQA's many exemptions apply. If so, no additional CEQA review is required. If not, the lead agency determines the type of review appropriate based on whether the project may have an adverse environmental impact. To make this finding, the agency prepares an **initial study**. If the initial study shows that the project will not have any adverse effects, the agency prepares a **negative declaration** describing the reasons the project will not harm the environment. On the other hand, if the initial study shows that the project might have adverse effects, the agency must prepare an **environmental impact report** (EIR) to analyze those impacts.

EIR = Environmental impact report

Does CEQA apply? The first task is for the lead agency to determine whether CEQA applies to the activity under consideration. This is subject to two important limits: (1) CEQA does not apply when it can be seen with certainty that the proposed activity will not have a significant effect on the environment, and (2) CEQA only applies to activities that constitute a **project**. The first restriction is self-explanatory, since CEQA is an environmental review statute. However, the provision limiting CEQA's application to a project requires understanding how this term is defined.

CEQA does not apply: (1) when it can be seen with certainty that the activity will not have a significant effect on the environment, or (2) when the activity does not constitute a project.

Under CEQA, a project is the whole of an action that can result in either a "direct physical change in the environment or a reasonably foreseeable indirect physical change in the environment...." Pub. Res. Code § 21065; CEQA Guidelines § 15378(a); *Simi Valley Rec. & Park Dist. v. Local Agency Formation Comm'n* (1975) 51 Cal. App. 3d 648, 663–664; *Friends of Mammoth v. Bd. of Supervisors* (1972) 8 Cal. 3d 247, 259–263. The CEQA Guidelines define "project" to include an activity involving the issuance to a person of a lease, permit, license, certificate, or other entitlement. *Id.* § 15378(a)(3); *Sierra Club v. County of Sonoma* (1992) 6 Cal. App. 4th 1307, 1315 (the definition of project "encompasses a wide spectrum, ranging from the adoption of a general plan...to activities with a more immediate impact, such as the issuance of a conditional use permit for a site-specific development proposal.") However, activities that involve purely ministerial actions in which the lead agency does not exercise its discretion are not "projects" that trigger CEQA review.

Under CEQA, a project is the whole of an action that has a potential for resulting in either a direct physical change or reasonably foreseeable indirect change in the environment.

Purely ministerial actions do not constitute projects to which CEQA applies. However, doubts as to whether an activity is a project should be resolved in favor of finding that it is and assuming CEQA applies.

CEQA should be interpreted to "afford the fullest possible protection to the environment within the reasonable scope of the statutory language." CEQA Guidelines

§ 15003(f); *see also Laurel Heights Improvement Ass'n v. Regents of Univ. of California* (1988) 47 Cal. 3d 376, 390; *Friends of Mammoth v. Bd. of Supervisors* (1972), 8 Cal. 3d 247, 259; *Azusa Land Reclamation Co. v. Main San Gabriel Basin Watermaster* (1997) 52 Cal. App. 4th 1165, 1193. Therefore, any doubts as to whether an activity constitutes a project subject to CEQA should be resolved in favor of a finding that the action at issue is a project.

When seeking multiple authorizations, the "project" is the overall activity being considered. In evaluating its environmental impacts, the initial study must consider all phases of the project. CEQA Guidelines § 15063. Likewise, although a single project may require several separate discretionary approvals by governmental agencies, each approval does not constitute a separate CEQA project. Rather, the impacts associated with all of the approvals must be considered as a whole. CEQA Guidelines § 15378(c); *Rio Vista Farm Bureau Center v. County of Solano* (1992) 5 Cal. App. 4th 351, 370, *quoting Bozung v. Local Agency Formation Comm'n* (1975) 13 Cal. 3d 263, 283–284 ("The requirements of CEQA cannot be avoided by piecemeal review which results from 'chopping a large project into many little ones—each with a minimal potential impact on the environment—which cumulatively may have disastrous consequences.'").

The project is the overall activity being considered. If the applicant is seeking multiple discretionary approvals, these should all be evaluated as part of the project.

As a practical matter, CEQA will almost always apply to new applications involving surface mining operations. On the other hand, CEQA might not apply to activities relating to an existing mining operation. For example, those operations may constitute part of the pre-CEQA environmental setting, or they may be already fully entitled and not subject to additional discretionary approvals by the lead agency. Before proceeding to prepare CEQA documents, the applicant and the lead agency's staff should carefully consider whether CEQA applies to the proposed activity.

Is the project wholly or partially exempt from CEQA review? If the activity qualifies as a project subject to CEQA, the lead agency must determine whether the proposed project qualifies for one of the categorical or statutory exemptions from CEQA requirements. *Friends of Sierra Madre v. City of Sierra* Madre (2001) 25 Cal. 4th 165, 185, *quoting Friends of "B" Street v. City of Hayward* (1980) 106 Cal. App. 3d 988, 999–1000. If so, the lead agency prepares and adopts a notice of exemption and terminates the environmental review process.

It is unlikely that a wholly new mining operation will qualify for an exemption. In some cases, though, activities that propose to modify an existing operation may be exempt. For example, modifications to an existing operation may qualify under the **existing facilities** exemption of CEQA Guidelines section 15301. The application of exemptions depends on the facts of the particular case. Applicants and lead agency staff should carefully consider the surrounding factual circumstances when determining whether a proposed project is exempt from review.

Does the project have the possibility of causing significant environmental impacts? When the activity is a proposed project and no exemption applies, the lead agency must complete the third step of the CEQA review process, determining the type of documents to prepare. The appropriate type of CEQA documentation depends on whether the project has the possibility of causing a significant impact on the environment. To aid in making this determination, the lead agency usually conducts an initial study. If the initial study reveals that the project will not have any significant environmental impacts, the lead agency should complete a negative declaration. On the other hand, if the project may have a significant effect on the environment, an EIR must be prepared.

If an agency determines a project is not exempt, it usually conducts an initial study to determine if the project may have a significant effect on the environment.

The first step in calculating any impacts the new project may have is to establish the environmental setting or **baseline** conditions. This step is particularly important when evaluating a mining operation. If the environmental baseline includes an existing operation, the impacts associated with modification may be minimal or even environmentally beneficial. In appropriate cases, any impacts associated with modification may be found to be insignificant. For this reason, the baseline is a key factor in determining impacts caused by a proposed project, and the appropriate CEQA document needed for review.

The environmental baseline is a key consideration in determining impacts a proposed new project might cause.

After establishing the baseline, an initial study should be prepared to evaluate the impacts of the proposed project compared to baseline conditions. A lead agency generally uses a version of the checklist reproduced in Appendix G of the CEQA Guidelines as a template for an initial study. However, staff must also include the agency's analysis and reasoning; it is not sufficient to check off the boxes that might represent the impacts of a project.

If the lead agency determines a project has the potential to cause a significant effect on the environment, it must prepare an EIR.

If the initial study reveals substantial evidence that an aspect of the project—individually or cumulatively—may have a significant effect on the environment, then the lead agency must prepare an EIR unless a previously prepared EIR has already disclosed its impacts. Pub. Res. Code § 21100; CEQA Guidelines § 15063(b)(1). A significant effect on the environment is a "substantial, or potentially substantial, adverse change in any of the physical conditions within the area affected by the project including land, air, water, minerals, flora, fauna, ambient noise, and objects of historic or aesthetic significance." CEQA Guidelines § 15382.

A negative declaration or mitigated negative declaration may be used when a project would not have a significant effect on the environment or mitigation would reduce such effects to below the threshold of significance.

When no substantial evidence reveals, in light of the whole record before the agency, that the project may have a significant effect on the environment, the lead agency should prepare a negative declaration. Pub. Res. Code § 21080(c); CEQA Guidelines § 15070(a). The most common type of negative declaration is a **mitigated negative declaration.** A mitigated negative declaration is prepared when the initial study identifies potentially significant environmental impacts, but the lead agency imposes mitigation measures to eliminate or to reduce those impacts below the threshold of significance. CEQA Guidelines § 15070(b)(1). A mitigated negative declaration describes how the proposed project will avoid or mitigate each of its potentially significant effects on the environment. Generally, the mitigation measures that the lead agency is relying on to reduce the identified environmental impacts should be included in the document before it is released for public review. If mitigation measures are changed or substituted after public review of the negative declaration, the document may need to be recirculated for another period of public review.

An EIR is required whenever substantial evidence indicates there is a fair argument a project might have a significant effect on the environment.

Even when mitigation measures are incorporated, it is difficult for a mining project to qualify for a negative declaration because the threshold for triggering preparation of an EIR is quite low. A lead agency must prepare an EIR whenever it can be fairly argued, on the basis of substantial evidence in the record, that the project may have a significant environmental impact. *Friends of "B" Street v. City of Hayward* (1980) 106 Cal. App. 3d 988, 1001; *No Oil, Inc. v. City of Los Angeles* (1974) 13 Cal. 3d 68, 75. In most cases, a new mining project will meet this low threshold and, thus, require preparation of an EIR. However, depending on the circumstances, a negative declaration may be appropriate for a proposed modification to an existing operation.

The degree of public controversy surrounding a proposed mining project may weigh in favor of an EIR instead of a negative declaration. The mere existence of serious public controversy surrounding the environmental effects of a project does not require preparation of an EIR. *Citizens for Responsible Dev. v. City of W. Hollywood* (1995) 39 Cal. App. 4th 490, 500 (public controversy does not constitute substantial evidence that project may have significant environmental effect); *Leonoff v. Monterey County Bd. of Supervisors* (1990) 222 Cal. App. 3d 1337, 1359 (absent any substantial evidence of potentially significant environmental effects, public controversy regarding the project did not preclude negative declaration); *Citizens Ass'n for Sensible Dev. v. County of Inyo* (1985) 172 Cal. App. 3d 151, 173 (public controversy alone does not require preparation of an EIR); *Perley v. Bd. of Supervisors* (1982) 137 Cal. App. 3d 424, 436.

PRACTICE TIP

When a project is controversial, it may be more efficient to prepare an EIR at the outset. Controversial projects generate most CEQA litigation. If a negative declaration is successfully challenged in court, the remedy ordered will be preparation of an EIR, but, due to the time it takes to litigate a matter, this order may not be made for many months or even years after the project was initially proposed. Thus, for a controversial project, it may actually be faster and cheaper to prepare an EIR than to attempt to sustain project approvals on a negative declaration.

Nonetheless, public controversy is a factor that the lead agency may take into account when determining whether to prepare an EIR or to proceed with a negative declaration. When there is serious public controversy about the impacts of the proposed project or their significance, it may signal that the agency's CEQA action will be subject to legal challenge. In such cases, it may be prudent for the lead agency to require preparation of an EIR.

A **first-tier EIR** is prepared when a lead agency is considering a new project. The four primary types of first-tier EIRs are: a **project EIR**, a **staged EIR**, a **program EIR**, and a **master EIR**. CEQA Guidelines §§ 15161, 15167, 15168, 15175. A project EIR, which analyzes a specific proposed project, is the most common. *Id.* § 15161. A project EIR is used for most mining operations. For some, however, a program or staged EIR is appropriate. A program EIR is used when a series of geographically related actions can be characterized as one large project, as parts in a chain of contemplated actions, or individual activities carried out under the same legal authority with generally similar environmental effects that can be mitigated in similar ways. *Id.* § 15168; *Sierra Club v. County of Sonoma* (1992) 6 Cal. App. 4th 1307, 1316 (county adopted a program EIR as part of a larger effort to regulate mining). A staged EIR is appropriate when a large capital project requires a number of discretionary approvals from government agencies and one approval will occur more than two years before construction begins. CEQA Guidelines § 15167.

A mining project that is not a wholly new proposal is not subject to first-tier environmental review. When earlier phases of a project have previously undergone environmental review, the agency may conduct subsequent-tier review for subsequently sought approvals. In this case, CEQA understandably provides for a more limited review process. In fact, once an EIR or negative declaration has been approved in conjunction with a project, CEQA limits the amount of additional review a lead agency can require for subsequent project approvals. The public agency cannot require preparation of another EIR for proposed revisions to, or later phases of, a previously analyzed project unless substantial evidence shows that new significant adverse impacts could result. Pub. Res. Code §§ 21166, 21167.2; CEQA Guidelines § 15162; *see also River Valley Pres. Project v. Metro. Transit Dev. Bd.* (1995) 37 Cal. App. 4th 154, 178; *Temecula Band of Luiseño Mission Indians v. Rancho Cal. Water Dist.* (1996) 43 Cal. App. 4th 425, 437.

When analyzing a later phase of a project or its proposed revision, the relevant question is whether the new incremental changes will have a significant environmental impact that was not analyzed in the previous review. *Temecula Band of Luiseño Mission Indians*, 43 Cal. App. 4th 437; *see also Benton v. Bd. of Supervisors* (1991) 226 Cal. App. 3d 1467, 1477. Thus, when the lead agency has previously certified an EIR or adopted a negative declaration for a mining operation, but the new project proposes changes to that previous approval, it may be appropriate for the lead agency to find that no additional review is required. Alternatively, the lead agency may require preparation of a **subsequent** or **supplemental EIR**, a **subsequent negative declaration**, or an **addendum.**

CEQA does not require a new EIR or negative declaration for subsequent phases of a project when the project has already been analyzed in an EIR or negative declaration.

A subsequent EIR is generally prepared for an already certified project when substantial changes are proposed, or new and important information—that was not known and could not have been known with the exercise of reasonable diligence at the time of certification—and its potential effects on the environment is discovered. Pub. Res. Code § 21166; CEQA Guidelines § 15162. The scope of a supplemental EIR is more limited than a subsequent EIR. A supplemental EIR is intended to make only those minor additions or necessary changes to make the previous EIR adequate. CEQA Guidelines, § 15163. A subsequent negative declaration is still a negative declaration and, consequently, should be used when it can be seen with certainty that the modifications to the project do not cause any significant environmental impacts. An addendum is appropriate when some changes or additions to the previous EIR or negative declaration are necessary, but the changes are not extensive enough to trigger preparation of a subsequent or supplemental EIR. *Id.* § 15164.

The appropriate type of subsequent-tier environmental document depends on the extent of the changes to the proposed project and the type of environmental document prepared for the first-tier review.

EIR Preparation and Review Process

Generally speaking, preparation of an EIR is expensive and time-consuming. An EIR can take a year or more to complete, with the cost exceeding $100,000. An applicant for a mining project should be prepared to pay for this expense, as most cities and counties require reimbursement for the cost of the environmental review for a private development project.

Essentially, the process has five steps: (1) scoping, (2) preparation of the **Draft EIR**, (3) circulation of the Draft EIR for public review, (4) preparation of responses to comments and issuance of the **Final EIR**, and (5) adoption of **findings**, a statement of overriding considerations (when necessary), a mitigation monitoring or reporting program, and approval of the project.

The first step in the preparation of an EIR is the scoping process, during which the lead agency solicits public comments regarding what issues should be analyzed in more depth in the EIR.

During the scoping process, the lead agency solicits comments on the EIR's scope, focus, and contents. Because a mining project may be subject to the jurisdiction of a state agency—or have the possibility of creating regional environmental impacts—the lead agency may be required to give notice of and conduct a meeting to afford interested persons an opportunity to comment on the scope of what the lead agency should be evaluating.

After scoping, a Draft EIR is prepared, a step that may take many months or even years.

Next, the lead agency reviews all comments received during the scoping process and begins to prepare the Draft EIR. Either the lead agency or the project applicant may hire a consultant for this step, but the lead agency retains discretion to control the document's contents, and the final product must reflect its independent judgment. Biological surveys, air quality modeling, geotechnical testing, and similar types of work may need to be performed before the lead agency can complete its analysis of the project's impacts. These preliminary studies may take several months or even years to complete.

In addition to the CEQA technical studies, the California Water Code requires an additional study for large projects proposing to use more water than required for 500 residential units. In this case, the entity supplying water for the proposed project must provide a "water supply assessment." *See* Water Code §§ 10910, *et seq*. This assessment serves as the basis for analyzing the project's impacts on the water supply and must be part of the EIR. Whether the requirement applies to a mining project is not entirely clear, but it's possible that water supply assessment laws could be interpreted to include projects such as these. As explained below, because only a very large mining operation would consume this much water, most mining projects will not trigger the requirement.

Once a Draft EIR is prepared, it is released for public review and comment.

When complete, the Draft EIR is released to the public for review and comment. The length of the review period often depends on whether the project will invoke the jurisdiction of a state agency or have the possibility of creating regional environmental impacts. Most mining projects meet one or both of these criteria, and the Draft EIR will usually have to be circulated for 45 days. However, an EIR prepared for a mining project that does not meet these criteria could be circulated for a shorter 30-day review period.

Although CEQA labels the document released for public review a Draft EIR, this is a misnomer. In fact, a Draft EIR contains the lead agency's CEQA analysis and constitutes a finalized document ready for public review. The document is labeled "Draft," because it does not yet incorporate the public comments or the lead agency's responses to the comments. Anyone may submit written or oral comments to the lead agency on the Draft EIR.

CEQA requires the lead agency to respond in writing to significant environmental issues raised in the comments submitted on the Draft EIR before the close of the public review period.

The Final EIR consists of comments received on the Draft EIR along with the lead agency's responses and any errata or other minor changes. (While CEQA allows a lead agency to make changes, substantial changes may trigger a need to recirculate the Draft EIR for another round of public review.) The lead agency is required to respond to all significant environmental issues raised in the comments received on the Draft EIR. Usually, an agency will take a few weeks after the close of the comment period to prepare its responses.

PRACTICE TIP

If the EIR determines that a project's significant environmental impacts cannot be mitigated, the lead agency must adopt a Statement of Overriding Considerations. When stating its reasons for approving a project despite its environmental impacts, the lead agency is not limited to environmental concerns. Social, economic, legal, and other factors, such as the need for local minerals, provision of local jobs, or generation of additional sales tax, may be used to support a Statement of Overriding Considerations.

While preparing the Final EIR, the lead agency is simultaneously drafting its findings for the project. The findings are intended to explain the lead agency's conclusions about the significance of the environmental impacts analyzed, the basis for the lead agency's selection of alternatives, and the rationale underlying each of the measures imposed on the project to mitigate its potentially significant impacts. When a project has significant impacts that cannot be avoided or mitigated, the lead agency will have to adopt a **Statement of Overriding Considerations**. *See* Pub. Res. Code § 21081(a)(3), (b); CEQA Guidelines §§ 15091(a)(3), 15093. The Statement of Overriding Considerations, which is intended to explain why the lead agency is willing to accept each significant impact the project will cause, is normally included in the lead agency's findings. Essentially, the Statement of Overriding Considerations reflects the lead agency's conclusion that identified social, economic, legal, technical, or other benefits of the project outweigh its significant environmental impacts.

At each step of the EIR process various notices must be given to the public and other affected agencies. Since CEQA is essentially a disclosure statute, courts have sometimes found that the failure to provide a required notice invalidates the process, requiring that a lead agency begin anew. *See, e.g., Fall River Wild Trout Foundation v. County of Shasta* (1999) 70 Cal. App. 4th 482, 491 (negative declaration set aside because county failed to give required notice to trustee agencies). The notices—including their timing and required contents—are highly technical, and legal counsel should be consulted when undertaking environmental review to ensure that all actions comply with CEQA's requirements.

CEQA Applied to Mining Uses

Every CEQA project, including surface mining, must be reviewed based on its own facts, taking into account the unique location, surroundings, environmental conditions, and other circumstances. Thus, it is not possible to state with any degree of specificity the issues an EIR must address. Based on prior cases and experience, however, the following issues commonly arise when an agency considers a project involving a surface mine:

- Type of mining project that generally requires an EIR
- Importance of an accurate project description generally, and, in the specific context of mining, the importance of distinguishing operational aspects from its reclamation plan
- Significance of the environmental setting or baseline in determining the environmental impacts of a proposed project
- Role project alternatives play in mining project review
- Environmental impacts frequently triggered by a surface mining project

Review of Projects Including New Mines and Expansion of Existing Mines

Two circumstances that will most likely trigger CEQA review for a mining project are: (1) a proposed new mining project seeks an operating permit, or (2) an existing project seeks to expand its operations, often through a use permit issued by the city or county in which the mine resides. *See supra* chapter 7 (describing procedures and standards governing the permitting of new surface mines). Because mining is an intensive use of land that can significantly affect the environment, a new mining project will generally require preparation of an EIR. A number of cases have rejected lead agencies' attempts to rely on a negative declaration instead of an EIR for the environmental review. *See Riverwatch v. County of San Diego* (1999) 76 Cal. App. 4th 1428; *Sierra Club v. County of Sonoma* (1992) 6 Cal. App. 4th 1307; *Oro Fino Gold Mining Corp. v. County of El Dorado* (1990) 225 Cal. App. 3d 872, 881.

Generally, an EIR is required when an operator seeks to commence a new mining operation.

Depending on the particular circumstances, an existing mining operation seeking a new use permit or modification of an existing use may need to prepare an EIR. For an operation that seeks to expand its operations or increase its output beyond what is currently permitted, an EIR is often necessary. *See, e.g., Fairview Neighbors v. County of Ventura* (1999) 70 Cal. App. 4th 238, 243 (EIR prepared for conditional use permit allowing the expansion of mining activities); *Sierra Club* (1992) 6 Cal. App. 4th 1321 (rejecting lead agency's use of a negative declaration in approving use permit application for expanded mining operations); *but see Perley v. Bd. of Supervisors* (1982) 137 Cal. App. 3d 424, 437 (affirming lead agency's adoption of a negative declaration for conditional use permit allowing surface mining).

The *Fairview Neighbors*, *Sierra Club*, and *Perley* decisions are useful examples of the differing approaches taken by lead agencies to comply with CEQA when seeking approval to begin a new mining operation or to expand an existing operation. These cases show that, in both situations, it is usually necessary to prepare an EIR instead of a negative declaration. Preparing a proper EIR will help provide the lead agency with a defensible record for its decision and the project applicant with documentation to show that the analysis required by CEQA has been conducted. *See* Pub. Res. Code § 21061 ("The purpose of an [EIR] is to provide public agencies and the public in general with detailed information about the effect which a proposed project is likely to have on the environment. . . ."); CEQA Guidelines § 15003(d) ("The EIR is to demonstrate to an apprehensive citizenry that the agency has, in fact, analyzed and considered the ecological implications of its action.")

Fairview Neighbors *provides an example of a situation in which an agency required an EIR for an expanded mining operation and that EIR withstood challenge in court.*

Fairview Neighbors v. County of Ventura. *Fairview Neighbors* (1999) 70 Cal. App. 4th 238, is a useful case for describing the benefits of preparing an EIR when an existing mining project seeks to expand its mine or increase production. In *Fairview Neighbors*, the lead agency prepared an EIR to analyze the expanded operation. At the request of the mine's operator, the EIR was revised and recirculated twice over a five-year period before finally being certified by the county. When later challenged in court, the EIR withstood scrutiny.

Fairview Neighbors *involved a mine operator's proposal to expand and increase the boundaries of a mining operation for a site which had been continuously mined since 1948.*

CUP = Conditional use permit

In *Fairview Neighbors*, residents living near the site of an existing mining operation alleged that the EIR prepared for its proposed increased production and expansion did not comply with CEQA. The proposed project sought to expand the boundaries of—and increase production from—a site that had been mined since 1948. *Id.* 240. In 1961, the county issued a conditional use permit (CUP) for mining on the site known as CUP 1328. *Id.* The county modified and renewed CUP 1328 in 1976 when a new owner purchased the mine. In 1976, the lead agency prepared an EIR for CUP 1328 that addressed, among other things, the number of truck trips the project could generate (810 per day) and the total amount of aggregate the project could remove (1,800,000 tons). *Id.* 240.

In 1993, a new owner purchased and began to operate the mine. *Fairview Neighbors*, 70 Cal. App. 4th 241. At the time of the purchase, an application was pending to modify CUP 1328 to expand the boundaries of—and increase production from—the mining operation. *Id.* In 1991, the lead agency had prepared an EIR to evaluate potential impacts that might result from the mine's proposed expansion and increased production. *Id.* At the request of the new operator, the lead agency revised and recirculated the EIR in 1994 and further revised and released it again in 1996. The lead agency certified the EIR in 1996, and plaintiffs subsequently sued, alleging that, among other things, the EIR did not adequately address the traffic, air pollution, noise, or biological impacts associated with the proposed mine. *Id.* 240, 241.

Rejecting these challenges to the EIR, the court held that the EIR was valid. The EIR had appropriately used the projected traffic impact caused by the mine operating at full capacity, as permitted under CUP 1328, as the baseline against which to measure the significance of the traffic impacts of the proposed expansion and increased output. *Fairview Neighbors*, 70 Cal. App. 4th 242–243. The court noted that "[d]iscussing the possible environmental effects of the project based on actual traffic counts would have been misleading and illusory under the facts here. The flow of traffic for a mining operation fluctuates considerably based on need, capacity and other factors." *Id.* 243 (citing *Hansen Bros. Enters., Inc. v. Bd. of Supervisors*, 12 Cal. 4th 533, 545–546). The court also stated that:

> The production previously permitted under CUP-1328 is the daily average of 810 truck trips. Indeed, in 1989, peak operation of the mine under CUP-1328 generated 837 daily truck trips. *A complete new EIR may not have been necessary here; a supplemental EIR, a narrowed EIR under the concept of "tiering" or a partial exemption may have been reasonable here.*
>
> *Id.*, citing *Gentry v. City of Murrieta* (1995) 36 Cal. App. 4th 1359, 1373–1374 (emphasis added)

Although *Fairview Neighbors* does not specify the applicable exemptions, the court may have been referring to the Class 1 categorical exemption set forth in section 15301 of the CEQA Guidelines. Section 15301 allows expansion of "existing facilities" when "the project involves *negligible or no expansion* of an existing use." (Emphasis added.) However, an operator seeking to increase production or expand the mine in more than a *de minimis* amount should not use this exemption. Moreover, an exemption should generally only be applied to the whole of a project. *See Ass'n for a Cleaner Env't v. Yosemite Cmty. Coll. Dist.* (2004) 116 Cal. App. 4th 629 (suggesting that it may be improper for the lead agency to rely on partial exemptions that cover only a portion of the project).

The *Fairview Neighbors* court also concluded that the EIR had adequately addressed potentially significant air pollution, noise, and biological impacts. Because the impacts had been identified and discussed, and were not mitigable, the court found that the EIR had properly adopted a Statement of Overriding Considerations. *Fairview Neighbors*, 70 Cal. App. 4th 244; *see also Laurel Heights Improvement Ass'n v. Regents of Univ. of California* (1993) 6 Cal. 4th 1112, 1124 (a lead agency may approve a project with significant, unmitigable impacts if a statement of overriding considerations regarding those impacts is adopted). The court also rejected the petitioners' claim that, by certifying the EIR, the lead agency had improperly deferred mitigation. *Fairview Neighbors*, 70 Cal. App. 4th 244–245 (distinguishing *Sundstrom v. County of Mendocino* (1988) 202 Cal. App. 3d 296 because the negative declaration in *Sundstrom* had improperly relied on future proposed mitigation studies to provide presumed mitigation).

Fairview Neighbors *demonstrates the problems that a mine operator may face when seeking to expand a long-existing mine after the surrounding area has been developed with housing. In a case such as this, the expansion may be controversial, and it may be prudent to prepare an EIR from the outset.*

Fairview Neighbors is representative of what many mining operations seeking to expand face today. These circumstances may have even existed for decades, but over time non-mining uses have developed in the surrounding area, and the neighbors oppose expansion. In cases such as this, when expansion may create environmental impacts that cannot feasibly be mitigated to the point of insignificance, preparation of an EIR may be required.

Sierra Club v. County of Sonoma. In contrast to *Fairview Neighbors*, the lead agency in *Sierra Club v. County of Sonoma* (1992) 6 Cal. App. 4th 1307, chose to

Sierra Club *suggests preparation of an EIR for the proposed expansion of a mining operation may be prudent.*

prepare a negative declaration, and the court rejected the lead agency's action. *Sierra Club* is thus instructive as to the potential pitfall created when an EIR is not prepared for a proposed expansion of an existing operation.

In *Sierra Club*, the mining project was to be sited on land designated for agricultural use under a previously prepared program EIR. In 1981, Sonoma County had certified a program EIR known as the Aggregate Resources Management Plan, or "ARM Plan." *Sierra Club*, 6 Cal. App. 4th 1313. The ARM Plan regulated the county's gravel and hardrock mining and provided a specific plan for its regulation. *Id.* The ARM Plan designated approximately 2,000 acres in the Middle Reach of the Russian River as eligible for terrace mining use permits. *Id.*

The operator in Sierra Club *sought to expand its mining operation to include land designated for agricultural use. Reasoning that nearly all of the environmental impacts associated with the project had already been adequately analyzed in a prior EIR, the county prepared only a negative declaration for the expansion project.*

After the ARM Plan was approved, several use permits were issued for terrace mining along the Russian River. The original operator received a permit in 1985 to mine approximately 50 acres in an area near Healdsburg known as the Grace Ranch. *Sierra Club*, 6 Cal. App. 4th 1314. In 1986, a new mining operator acquired the Grace Ranch property and other assets from the original owner. In 1989, the new operator applied to amend the ARM Plan by changing 145 acres designated as agricultural under the Plan to "Managed Resource: Mineral" that would allow the land to be mined. *Id.* Concurrent with the proposed ARM Plan amendment, the new operator sought both a use permit to terrace mine 50 of the 145 acres that were the subject of the application to amend the Plan and another ARM Plan amendment to allow reclamation by refilling pits with processing sediments and other earth materials, rather than by the river diversion process.

The Board of Supervisors concluded that, except for the use of non-native earth fill for reclamation, all but one of the environmental impacts that might result from the proposed changes had already been considered in the ARM Plan EIR. *Sierra Club*, 6 Cal. App. 4th 1314. On that basis, the lead agency adopted a negative declaration and approved the application.

The Sierra Club petitioned for a writ of mandate, alleging, among other things, that the county violated CEQA by adopting a negative declaration and by not preparing a new EIR. The trial court granted the petition, finding "substantial evidence in the record supporting a fair argument that the proposed method of reclamation may have significant environmental effects. . . ." *Sierra Club*, 6 Cal. App. 4th 1314. The court ordered the lead agency to set aside its approval of the project and required preparation of an EIR before any further approval. *Id.* 1314–15. The mining operator appealed.

After reviewing the different types of EIRs available under CEQA, the court focused on the standard of review that should be applied to the lead agency's decision not to prepare a new EIR for the proposed terrace mining project. The operator argued that the standard from Public Resources Code section 21166 applied. *Id.* 1318. Section 21166 prohibits an agency from requiring a subsequent or supplemental EIR unless substantial changes are proposed either in the project or in circumstances that will require major revisions in the EIR, or unless certain new information has become available after certification of the prior EIR. *Id.* 1317, citing CEQA Guidelines § 15162. The court stated, "Section 21166 comes into play precisely because in-depth review has already occurred, the time for challenging the sufficiency of the original EIR has long since expired, and the question is whether circumstances have *changed* enough to justify *repeating* a substantial portion of the process." *Id.*, citing *Bowman v. City of Petaluma* (1986) 185 Cal. App. 3d 1065, 1072–1073, *and Long Beach Sav. & Loan Ass'n v. Long Beach Redevelop. Agency* (1986) 188 Cal. App. 3d 249, 265.

The court rejected the new mining operator's argument. Instead, the court stated that "section 21166 and its companion section of the Guidelines appear to control only when the question is whether more than one EIR must be prepared for what is essentially the same project." *Sierra Club*, 6 Cal. App. 4th 1320, citing *Fund for Envtl. Def. v. County of Orange* (1988) 204 Cal. App. 3d 1538, 1542–1545, 1552. The court then held:

> [The operator] concedes that if a proposed new activity is a separate project, the "fair argument" test should apply to an agency's decision whether to require a tiered EIR for the later project. [The operator] argues that here, its proposed activity was not a separate project, but was instead a part of or a minor modification of the single large project already studied in the ARM Plan. We cannot agree, as [the operator] sought permission to engage in terrace mining on land which was specifically designated in the Plan as an agricultural resource. [The operator] attempts to minimize the significance of that designation by asserting that it was based on nothing more than then existing ownerships; according to [the operator], those areas owned by mining companies when the Plan was prepared were designated as a mineral resource, and the remaining areas were designated for agricultural use. But that claim simply ignores the Plan itself, which states that lands in the "Managed Resource: Agriculture" category were "proposed for preservation for their value as both an agricultural resource and as groundwater recharge." Under these circumstances, the evidence does not support a determination that [the operator's] proposed site-specific project was either the same as or within the scope of the project, program, or plan described in the program EIR. (*See* Guidelines, § 15168, subd. (c)(5).) Therefore, section 21166 was inapplicable, and the County was obligated by section 21094, subdivision (c), to consider whether [the operator's] site-specific new project might cause significant effects on the environment that were not examined in the prior program EIR. If there was substantial evidence in the record that the project may arguably have such effects, the County should have required preparation of a [subsequent] tier[] EIR, notwithstanding the existence of contrary evidence, including contrary expert opinions.
>
> *Id.* 1321

Notably, the *Sierra Club* court took pains to distinguish the facts of that case from the situation where the proposed new activity is a modification of a previously approved project. This distinction has important consequences for the CEQA

The court in Sierra Club *concluded that the proposed expansion was a separate project from that which the lead agency had reviewed previously. It therefore invalidated the county's reliance on a negative declaration for the project.*

process because these cases are subject to a standard of review that is more deferential to the lead agency's discretion than the fair argument standard applied in *Sierra Club*. When the new application seeks to modify a previously approved project and the project has already undergone environmental review, Public Resources Code section 21166 applies. As long as substantial evidence is in the record to support the lead agency's decision not to require a supplemental or subsequent EIR, it must be upheld. *Friends of Davis v. City of Davis* (2000) 83 Cal. App. 1004, 1017, 1019; *Benton v. Board of Supervisors* (1991) 226 Cal. App. 3d 1467, 1477; *see also* Pub. Res. Code § 21166; CEQA Guidelines § 15162. Relying on this deferential standard, the court in *Benton* approved the use of a negative declaration to expand a previously approved 856-acre winery project by 120 acres, even though the expansion relocated the winery buildings one mile to the west of the originally proposed site, placing them closer to existing residences and changing the access routes to the redesigned site. *Benton*, 226 Cal. App. 3d 1473.

Sierra Club *makes clear that mine expansions not analyzed in previous environmental documents require a subsequent CEQA document. If the expansion constitutes a modification to a previously approved project, the CEQA document should focus on the incremental environmental changes that result from that modification.*

Like *Fairview Neighbors*, *Sierra Club* provides important guidance for mining operations seeking to expand their mines or increase production. First, assuming that the existing mining is operating under a certified EIR, the mine operator should carefully review the EIR to determine whether the proposed expansion or increased production was contemplated and addressed in the original EIR. Second, if the proposed expansion or increased production was not analyzed in the prior EIR, then the lead agency must determine whether the new application is properly characterized as a modification of the previously approved project or whether it constitutes a separate, new project. When the new application is a modification to a previously approved project, the lead agency's CEQA analysis should focus on the incremental impacts of the changes to the project. The lead agency must then determine, under Public Resources Code section 21166 and CEQA Guidelines sections 15162 and 15163, what additional documentation might be required. *See* Pub. Res. Code § 21068.5; CEQA Guidelines § 15152; *Sierra Club*, 6 Cal. App. 4th 1318 ("CEQA directs agencies to 'tier' EIR's whenever feasible. . . ."); *Las Virgenes Homeowners Fed'n, Inc. v. County of Los Angeles* (1986) 177 Cal. App. 3d 300, 307. In many cases, though, the proposed expansion or increased production could itself create significant environmental impacts, which would require preparation of a Subsequent or Supplemental EIR.

Perley *is a rare case in which a negative declaration was upheld for a proposed mining project. Its precedential value is questionable, however.*

Perley v. Board of Supervisors. The *Perley* case (1982) 137 Cal. App. 3d 424, 437 provides a relatively rare example of a proposed new mining project being approved on a negative declaration. However, a key aspect of *Perley's* holding—that the petitioner failed to present substantial evidence to support its claims—was decided before CEQA's 1993 codification of the definition of "substantial evidence." Subsequent cases have distinguished *Perley* on this basis, and a lead agency should be cautious in relying on it.

In *Perley*, a proposed mine operator applied for a zoning change and conditional use permit to allow it to mine talc. *Perley*, 137 Cal. App. 3d 429. The project proposed a wholly new mining operation. The county planning director asked the operator to obtain independent studies of the noise, dust, and traffic impacts of the proposed project, and to conduct meetings regarding the project with its neighbors and interested members of the public. *Id.*

After receiving the requested reports on noise, dust, and traffic from the proposed operator, the lead agency first recommended that the use permit be granted and a negative declaration adopted, subject to certain conditions of approval that would mitigate to the point of insignificance the environmental concerns (dust, noise, and water quality impacts) identified in the reports. *Id.* 429. However, after further public input concerning the project, the county planning commission voted to require preparation of an EIR because it found that the proposed project: (1) might have a significant effect on the environment, particularly as to groundwater, noise, dust, and roads, and (2) was the subject of serious public controversy. *Id.* The operator appealed the planning commission's decision to the Board of Supervisors, which found that—despite the concerns of nearby residents regarding increased traffic, noise, dust, and possible groundwater depletion or degradation—the proposed project had been revised and amended by the conditions of approval to satisfactorily mitigate any significant impacts upon the environment. The board adopted a negative declaration and approved the application for a use permit and the mine reclamation plan subject to the conditions proposed by the planning department. The plaintiff petitioned for a writ of mandate seeking to vacate the lead agency's approval of the project.

The court of appeal affirmed the superior court's denial of the petition for writ of mandate. The court of appeal held that the record contained no substantial evidence to support a fair argument that the proposed project would cause significant environmental effects. *Id.* 433–34.

In many respects, *Perley* is an unusual decision that is unlikely to be repeated today. First, it is rare for any new mining operation to be approved under a negative declaration (as opposed to an EIR), due to the intensity of most mining land uses and increased public sensitivity to environmental concerns. Second, while the lead agency in *Perley* focused its study of the proposed project's potential environmental impacts on noise, dust, and traffic, and the project opponents did not appear to challenge the scope of the county's environmental review, CEQA requires consideration of all of the project's potentially significant environmental impacts. *Protect the Historic Amador Waterways v. Amador Water Agency* (2004) 116 Cal. App. 4th 1099, 1110–11. Appendix G of the CEQA Guidelines suggests that, in conducting its initial study, the lead agency should consider environmental factors such as aesthetics, biological resources, agricultural resources, geology/soils, and recreation. Many mining projects have the potential to create significant environmental impacts in one or more of these categories.

PRACTICE TIP

The definition of "substantial evidence" codified after the *Perley* decision and subsequent CEQA case law make it inadvisable to rely on *Perley*. Today, if a planning commission makes findings that a negative declaration is not appropriate, it will be difficult to defend the negative declaration against any subsequent legal challenge.

Third, in upholding the lead agency's decision to prepare a negative declaration instead of an EIR, the court focused on whether the record contained substantial evidence showing that the project would cause significant environmental impacts. Subsequent to the *Perley* decision, CEQA's definition of "substantial evidence" was codified in 1993. Public Resources Code section 21082.2(c) now defines substantial evidence to include "facts, reasonable assumptions predicated upon facts, and expert opinion supported by facts." Given this definition, it is likely that the planning commission's vote in *Perley* (and supporting justifications for the vote) to require preparation of an EIR would constitute substantial evidence to support a fair argument that the project could have a significant environmental impact. *See Pocket Protectors v. City of Sacramento* (2004) 124 Cal. App. 4th 903 (among other things, planning commission's findings of fact that the project conflicted with applicable land use restrictions constituted substantial evidence to support a finding that the project might have a significant environmental impact); *see also Oro Fino Gold Mining Corp. v. County of El Dorado* (1990) 225 Cal. App. 3d 872, 882 (relevant personal observations can constitute substantial evidence under CEQA).

Although *Perley* remains an example of a proposed new mining operation that successfully used a negative declaration to satisfy CEQA, lead agencies and mining operators seeking to permit a new project or expand or increase production from an existing project should carefully assess whether to use a negative declaration. When used appropriately, a negative declaration or mitigated negative declaration can save time and money. However, when used without well-documented justification, a negative declaration becomes a liability that, in the long run, may be more expensive and time-consuming than preparing an EIR in the first instance.

Mining Project EIR Must Accurately Describe Project and Objectives—Differences Between Mining Operation and Its Reclamation Plan

The project description is the foundation upon which an EIR is built.

An accurate project description is the foundation upon which an EIR is built. The project description serves as a blueprint of sorts for the EIR process. As the court stated in *County of Inyo v. City of Los Angeles* (1977), "[a]n accurate, stable and finite project description is the *sine qua non* of an informative and legally sufficient EIR." 71 Cal. App. 3d 185, 193. The descriptions of a mining project should thus take care to describe the boundaries of the project and state its objectives and anticipated maximum output. Moreover, it may be necessary that the project description explicitly identify the difference between a mine's operations and its reclamation plan.

The boundaries of a mining project affect the scope of potential onsite environmental impacts as well as the physical boundaries within which mining may occur. Aside from practical considerations such as ingress and egress from the mining site and placement of mining equipment and related appurtenances, the project applicant should, prior to seeking permits, carefully survey the proposed site to identify sensitive habitat (*e.g.*, streams, drainages, and wetlands) to be avoided or, if not avoidable, addressed in project planning. For example, buffer zones and drainage infrastructure required to avoid impacts to sensitive habitats or species should be considered. Expansion of the mining site is likely to trigger further environmental review. *Fairview Neighbors v. County of Ventura* (1999) 70 Cal. App. 4th 238, 241 (mining operator's application to expand mining operation beyond boundaries of project site necessitated preparation of an EIR); *Dry Creek Citizens Coalition v. County of Tulare* (1999) 70 Cal. App. 4th 20, 24 (same). To avoid repeated time-consuming and expensive environmental review, the project applicant should attempt to ensure that the project site is large enough to accommodate reasonably foreseeable operation and production needs over time.

PRACTICE TIP

The project description should include reasonably foreseeable future expansions of the mining operations. Including these expansions in environmental review at an early stage will streamline subsequent phases of the CEQA process and facilitate subsequent project approvals.

The CEQA Guidelines require that the EIR state all objectives of the proposed project. CEQA Guidelines § 15124(b). This gives notice to the public, and helps an agency identify and evaluate the feasibility of alternatives. As discussed below, the lead agency may reject alternatives that cannot reasonably attain the objectives. Consequently, the applicant should carefully consider, and the description should clearly identify, the project's objectives.

The EIR must clearly identify the objectives of the proposed mining project.

A lead agency should also avoid drafting project objectives too narrowly. Doing so may cause the EIR's analysis of alternatives to be inadequate. *Kings County Farm Bureau v. City of Hanford* (1990) 221 Cal. App. 3d 692, 737–39; *City of Santee v. County of San Diego* (1989) 214 Cal. App. 3d 1438, 1454–55. If the objectives are drawn so narrowly that the project description does not allow for a meaningful discussion of alternatives, then the EIR may be subject to challenge.

In some cases, the project description may also need to draw a clear distinction between mine *reclamation* and mine *operations*. A permit for a new mining operation and a reclamation plan application may each be considered a CEQA project. *See El Dorado County Taxpayers for Quality Growth v. County of El Dorado* (2004) 122 Cal. App. 4th 1591, 1596–1597; *City of Ukiah v. County of Mendocino* (1987) 196 Cal. App. 3d 47, 54. However, they are not always part of the *same* project. As discussed in chapter 6, every surface mining operation, even one under "vested rights" as a nonconforming use, must receive approval for its reclamation plan. Pub. Res. Code § 2770(a); *El Dorado County Taxpayers*, 122 Cal. App. 4th 1600. A reclamation plan, however, does not constitute a permit to mine. Instead, it sets forth how an operator will reclaim mined land by preparing the property for post-mining use and remediating adverse conditions that remain after mining is completed. While a permit governs how a surface mine operates, a reclamation plan ensures that the aesthetic, public safety, and environmental costs of unreclaimed property are not forced on the public, but are instead borne by the mining operator. *People ex rel. Dep't of Conserv. v. El Dorado County* (2005) 36 Cal. 4th 971, 990–991. Thus, the adverse environmental impacts of a reclamation plan—if any—are very different from the impacts associated with mining. When an operator seeks approval of only a reclamation plan or plan amendment, the project may have few if any adverse environmental consequences.

PRACTICE TIP

Because agencies may reject project alternatives that would not feasibly attain the project's objectives, an applicant should take great care in drafting realistic objectives that explain the key purposes of the project. Carefully drafted objectives can then be used to reject project alternatives that do not meet these aims. For example, when an aggregate mine is proposed, the shortage issues discussed in chapter 2 may provide a significant benefit to justify a Statement of Overriding Considerations that will help a lead agency reject alternatives that do not increase aggregate supplies.

Opponents sometimes attack reclamation plans by attempting to blur the distinction between the plan and the operation. In *El Dorado County Taxpayers*, for instance, the county approved a reclamation plan for a mining operation by adopting a negative declaration after determining that the proposed reclamation plan would not have a significant effect on the environment. *Id.* 1595. The applicant did not seek a permit to mine, but sought only to amend its reclamation plan at that time. The petitioners challenged the county's decision, arguing that the county should have prepared an EIR instead of a negative declaration. Among other things, the petitioners alleged that "the initial study and negative declaration improperly analyze only the reclamation aspect of the project divorced from its mining aspect...." *Id.*

The court rejected the petitioners' claims, concluding that plaintiffs had failed to distinguish between the environmental impacts from mining activity and the impacts from reclamation plan activity. *El Dorado County Taxpayers*, 122 Cal. App. 4th 1595. Specifically, the court stated that, "[a]lthough mining reclamation is not done without there having been mining, *the project at issue presented the environmental effect of a reclamation plan rather than the environmental effect of mining activity*." *Id.* 1598 (emphasis in original). On this basis, the court of appeals upheld the lead agency's conclusions that the reclamation plan would not have a significant effect on the environment and could be approved on a negative declaration. *Id.* 1604.

PRACTICE TIP

When seeking approval or amendment of a reclamation plan for an existing legal operation, a lead agency should normally analyze only impacts attributable to the reclamation plan. This is particularly true when the operation is a legal nonconforming use. An operator allowing environmental documents prepared for a reclamation plan to address unrelated operational issues—noise, dust, hours of operation, and the like—risks opening the project up to "back door" regulation.

If non-reclamation issues are made part of the environmental review, new mitigation measures will likely be imposed, and approval of the reclamation plan might lead to imposition of new conditions. Expanding CEQA review in this manner is not appropriate unless an applicant has sought the agency's approval for some change in its operations.

City of Ukiah is another example of the importance of distinguishing between the operation and its reclamation plan. In that case, a mining operation established in 1946 sought approval for its reclamation plan after enactment of the Surface Mining and Reclamation Act and Mendocino County's adoption of a mining ordinance. *Id.* 49–50. Even though the mine operator only sought approval of a reclamation plan, the county treated it as an application for a use permit authorizing gravel extraction. *Id.* 50. Planning department staff recommended preparation of an EIR or an "equivalent hydrologic study." *Id.* 51. Eventually, the county accepted the position that the operator had a legal, vested right to extract gravel. Once mining was removed from the project's environmental analysis, the county adopted a negative declaration for the reclamation plan. *Id.*

The City of Ukiah filed suit, seeking a writ of mandate to set aside the negative declaration and compel preparation of an EIR. *Id.* Both the trial court and the court of appeal denied relief, finding that the administrative record adequately supported the county's determination that the project—consisting of only the reclamation plan—would not cause significant environmental effects. *Id.* 55 ("The crucial consideration herein was the limited nature of the proceeding: whether [the operator's] reclamation plan should be approved.").

A description that clearly distinguishes between the mining project and its reclamation plan will help both a lead agency and the public differentiate the two types of projects. Such clarity also serves a project applicant by allowing for more efficient agency review and the potential for fewer CEQA-related challenges. *Id.* 54

("Considerable confusion reigned at the board hearing as to the nature of the project then under review.").

EIRs Must Describe the Project's Environmental Setting

Accurately describing the environmental setting is important for any proposed project, but this CEQA requirement can be particularly important for surface mining. The environmental setting or baseline serves as the measuring point against which the proposed project's potential environmental effects will be calculated to determine their significance. CEQA Guidelines § 15125(a); *Friends of the Eel River v. Sonoma County Water Agency* (2003) 108 Cal. App. 4th 859, 874; *Save Our Peninsula Comm. v. Monterey County Bd. of Supervisors* (2001) 87 Cal. App. 4th 99, 120–121; *County of Amador v. El Dorado County Water Agency* (1999) 76 Cal. App. 4th 931, 952. Normally, the baseline is described as the physical environmental conditions in the vicinity of the project at the time the lead agency publishes its notice (**notice of preparation**) that the agency intends to prepare an EIR. CEQA Guidelines § 15125(a); *Save Our Peninsula*, 87 Cal. App. 4th 121 ("As various courts, including this one, have held, the impacts of the project must be measured against the real conditions on the ground.").

CEQA's requirement to accurately describe a project's environmental setting can be particularly important for a mining project.

Accurately describing the CEQA baseline is particularly important for surface mining. First, an EIR that seeks to expand mining or increase production should clearly state that the project is onsite or adjacent to an existing operation. Highly disturbed land associated with an existing operation may allow for more efficient review of some environmental impact categories but require more extensive review of others. For example, endangered or threatened species do not likely reside within the boundaries of an ongoing operation, but a mine that seeks to expand its boundaries or increase its production may require a more robust cumulative impacts analysis of noise, air quality, and hydrology/water quality, and similar topics. *See* CEQA Guidelines § 15130; Pub. Res. Code § 21100. Expansion of the mine may in fact add to already significant environmental impacts. *See, e.g., Fairview Neighbors v. County of Ventura* (1999) 70 Cal. App. 4th 238, 244–246 (lead agency prepared "separate lengthy section of the EIR devoted to analysis of the cumulative impacts" of the expansion of an existing mine and concluded that the project would cause unmitigable noise and air quality impacts).

EIRs that seek to expand mining operations or increase production should clearly describe that the project is on or adjacent to existing mining operations.

Second, the description for a new operation proposed on land already highly disturbed should clearly reflect that fact. It is not uncommon for a new mining operation to be proposed for land in close proximity to an existing industrial use (*see Fairview Neighbors*, 70 Cal. App. 4th 238 (proposed project site already occupied by industrial facilities used to make concrete batch, road base, and mortar)) or on agricultural land (*see Sierra Club v. County of Sonoma* (1992) 6 Cal. App. 4th 1307, 1314 (new mining project proposed for lands designated as agricultural uses)). Because a mining project is sometimes located on land currently or formerly used

Where new mining operations are proposed on land that is already highly disturbed, the project description should clearly reflect that fact.

for industrial purposes, the CEQA baseline should reflect any significant known quantities of contaminants or chemicals present at the site that may be affected by the proposed project. Identifying these substances will provide notice to the public and help insulate the project EIR from attack. *Baird v. County of Contra Costa* (1995) 32 Cal. App. 4th 1464, 1468 (rejecting challenge to project EIR that alleged building the project on potentially contaminated site was a significant effect under CEQA: "The purpose of CEQA is to protect the environment from proposed projects, not to protect proposed projects from the existing environment.").

Third, some mining projects are located in environmentally sensitive areas that should be carefully described. It is not uncommon for an aggregate mine to be located adjacent to a river that deposits gravel along its banks during high flows. *See, e.g., City of Ukiah v. County of Mendocino* (1987) 196 Cal. App. 3d 47, 54 (mining operator extracted gravel from riverbed that is replenished during high-flow events). The environmental documents for these projects should examine the potential impacts on the stream, its fisheries, riparian habitat, and the like.

Finally, a mining project proposed to be in close proximity to a residential or other urban land use should take particular care to establish the CEQA baseline for aesthetics, noise, air quality, and traffic. Projects proposed near a residential area with very little or very high noise levels, degraded air quality, or existing traffic problems are likely to trigger significant scrutiny from the lead agency and the public. Even an incrementally minor project-caused impact can be significant when added to an area with already degraded conditions. *Berkeley Keep Jets Over the Bay Comm. v. Bd. of Port Comm'rs* (2001) 91 Cal. App. 4th 1344, 1380–1381 (compliance with noise ordinance may not be determinative as to issue of significant impact under CEQA if project noise is perceived to be significant). Conversely, a mining project proposed for an area without degraded conditions can be perceived by those living near the site as having an impact that is significant. *Oro Fino Gold Mining Corp. v. County of El Dorado* (1990) 225 Cal. App. 3d 872, 882 (relevant personal observations can constitute substantial evidence of environmental impacts).

PRACTICE TIP

Some courts view the personal observations of a project's neighbors as substantial evidence of potential environmental impacts. Cases addressing this issue are mixed, and it is often not possible to predict how future courts will interpret the body of law. Consequently, it may be prudent to assume that the statements of neighbors opposed to the project may be deemed to be substantial evidence of significant environmental impacts, particularly when the personal observations relate to the project's anticipated aesthetic, noise, or traffic effects. If this evidence is presented to the lead agency when the lead agency is considering a negative declaration, it may be wise to re-evaluate whether an EIR is warranted.

EIRs Should Adequately Describe Potential Alternatives to the Proposed Project

CEQA requires that an EIR analyze potential alternatives to the proposed project to help determine whether significant, avoidable environmental damage that might otherwise be caused by the project may be reduced or eliminated through the adoption of an alternative. The alternatives analysis helps effect the basic purposes of CEQA and is an important aspect of an EIR. Pub. Res. Code § 21002; CEQA Guidelines § 15002(a)(3).

The CEQA Guidelines provide that an EIR must describe a reasonable range of feasible alternatives to the proposed project:

> An EIR shall describe a range of reasonable alternatives to the project, or to the location of the project, which would feasibly attain most of the basic objectives of the project but would avoid or substantially lessen any of the significant effects of the project, and evaluate the comparative merits of the alternatives. An EIR need not consider every conceivable alternative to a project. Rather it must consider a reasonable range of potentially feasible alternatives that will foster informed decision making and public participation. An EIR is not required to consider alternatives which are infeasible. The lead agency is responsible for selecting a range of project alternatives for examination and must publicly disclose its reasoning for selecting those alternatives. There is no ironclad rule governing the nature or scope of the alternatives to be discussed other than the rule of reason.
>
> CEQA Guidelines § 15126.6(a). CEQA Guidelines section 15364 defines feasible as "capable of being accomplished in a successful manner within a reasonable period of time, taking into account economic, environmental, legal, social, and technological factors."

CEQA requires that an EIR consider a reasonable range of feasible alternatives to the proposed project.

An EIR is required to identify and analyze only those alternatives necessary to allow for a reasoned choice between the proposed project and the proposed alternatives. This reasoned choice standard is called the "rule of reason." *Citizens of Goleta Valley v. Bd. of Supervisors* (1990) 52 Cal. 3d 553, 565. The range of potential alternatives must include those that could feasibly accomplish most of the project's basic objectives and could avoid or substantially lessen one or more of the significant effects. The EIR should briefly describe the rationale for selecting the alternatives to be discussed and identify any that were considered by the lead agency and rejected as infeasible during the scoping process. CEQA Guidelines § 15126.6(c).

Agencies apply a rule of reason in determining the number and range of alternatives to consider.

Project alternatives are often hotly contested. It is not unusual for opponents to demand that a proposed project be modified or an alternative to the project be adopted to accommodate concerns related to its potential impacts on traffic, noise, aesthetics, or biological resources on and near the project site. Opponents may argue for the "no project" alternative; a reduced intensity alternative, usually consisting of a smaller project at the same site; or an alternative location.

While an EIR must describe a range of reasonable alternatives, the alternatives must be capable of feasibly attaining most of the project's basic objectives. CEQA Guidelines § 15126.6(a). Thus, alternatives that cannot feasibly obtain most of the basic objectives may be rejected. *Sequoyah Hills Homeowners Ass'n v. City of Oakland,* 23 Cal. App. 4th 704, 715 (upholding the city's rejection of one of the proposed alternatives on the ground that it "would defeat the project objective of providing 'the least expensive single-family housing for the vicinity.'").

Project alternatives that will not feasibly attain most of the basic project objectives may be rejected.

As discussed in chapter 1, a surface mine cannot be located just anywhere. Finding a suitable site with adequate resources is a lot like finding a needle in a haystack, and relying on another location as a viable alternative may not be possible. This is especially true if project site is a state-designated significant mineral resource, described in chapter 4. To constitute a true alternative, an offsite location may have to

Generally, it will be difficult to find suitable offsite alternative locations for proposed mining projects. The lead agency will often be able to reject consideration of offsite alternatives early on in the CEQA process.

be designated a significant mineral resource or provide a similar grade and extent of the resource the applicant seeks to mine. For most projects, therefore, viable project alternatives will be limited to alternatives onsite, providing for a different size, mining plan, or duration.

For most mining projects, viable project alternatives would likely be limited to onsite alternatives.

Environmental Impacts Often Associated with Surface Mining

Because mining is an intensive use of land occurring on many acres that continues in stages for decades, uses large amounts of water, and generates noise and dust on an ongoing basis, a variety of environmental impacts are created. CEQA documents should carefully assess whether the proposed project may significantly affect: (1) water supply, (2) water quality associated with discharges or potential discharges, (3) air quality, (4) biological resources (including both terrestrial and aquatic species), (5) aesthetics, and (6) noise and vibration. Each of these deserve attention, but CEQA case law suggests that water supply, water quality, and air quality are the most frequently litigated CEQA impacts associated with a mining project.

CEQA case law suggests that water supply, water quality, and air quality are the most frequently litigated CEQA topics associated with a mining project.

While some environmental aspects of mining—including regulation related to water and air quality—are covered in chapter 7, this section addresses why an EIR should carefully address possible impacts a proposed project may have on water supply, water quality, and air quality. In fact, a new project almost always affects these categories. Water is scarce and increasingly regulated in California, and in many regions air quality is poor. The incremental addition of air pollution caused by mining is frequently deemed a significant impact under CEQA.

Water supply and quality impacts. When washing processes are involved, mining requires large quantities of water. Therefore a reliable supply is essential, and a mining operation must employ measures that appropriately dispose of its **process water**. Environmental documents typically address both of these aspects of mining.

New legislation and related case law emphasize the importance of specifying whether an adequate water supply will be available for proposed projects.

Water supply. While CEQA case law has long required that environmental documents sufficiently describe the source of water for a project, 2002 legislation and related case law emphasize the importance of specifying whether an adequate supply will be available for a project. In *Santiago County Water Dist. v. County of Orange* (1981) 118 Cal. App. 3d 818, a water district challenged the sufficiency of an EIR prepared to analyze a proposed sand and gravel mining plant. The court found that information about the project's water supply was lacking in two significant ways. First, the court found that the EIR did not discuss facilities that would have to be constructed for its delivery to the operation. *Id.* 829. Second, the court determined that the EIR failed to adequately address the environmental impact of supplying the large quantities of water that the project would require. *Id.* 830. Critical of the lead agency's unsupported representations that sufficient water existed to serve the project, the court stated:

The county has attempted to remedy the inadequacies of the EIR by presenting evidence to the trial court to show that there are sufficient water resources available for the project. Indeed, the trial court made findings of fact to such effect. This, however, is beside the point. It is the adequacy of the EIR with which we are concerned, not the propriety of the board of supervisors' decision to approve the project. "[Whatever] is required to be considered in an EIR must be in that formal report; what any official might have known from other writings or oral presentations cannot supply what is lacking in the report."

Id. 831 (citing *Environmental Defense Fund, Inc. v. Coastside County Water Dist.* (1972) 27 Cal. App. 3d 695, 706)

The Santiago County Water District *case emphasizes the need for EIRs to clearly identify how a mining operation would meet its water supply needs. If the project requires the development of new water supply infrastructure, those facilities need to be analyzed in the environmental document.*

Santiago County Water District underscores the importance of addressing in the environmental document any potentially significant impacts that may be caused by a proposed project's use of water, whether stemming from the infrastructure required for delivery or otherwise.

A mining operation that would require a water supply equivalent to a 500-dwelling unit project must comply with the water assessment requirements of SB 610.

Furthermore, in 2002, California enacted Senate Bill 610, which amended sections of the Public Resources Code and Water Code. Among other things, SB 610 requires that, for certain development projects subject to CEQA, cities and counties must consider water supply assessments prepared for the proposed development to determine whether projected water supplies available to the proposed project will meet the project's anticipated demand. Water Code § 10910; Pub. Res. Code § 21151.9. Projects subject to this requirement include projects that would have a water demand equal to or greater than a 500-dwelling unit project. Water Code § 10912(a)(7).

It is unlikely, but possible, that an extremely large mining operation could use enough water to trigger compliance with the SB 610's water assessment and related requirements. The average 500-dwelling unit project in California will annually use several million gallons of water. American Water Works Association, *Fact Sheets: Residential Uses of Water Report* (Dec. 2005) ("The North American households included in this study use approximately 146,000 gallons annually"). Although water use for mining operations can be highly variable depending on mine type, it is possible that an extremely large gravel and aggregate mining operation could annually consume as much a 500-dwelling unit project. Given this, it may be wise to obtain a water supply assessment from the entity that will serve water to the proposed mining project when preparing an EIR for a large-scale mining operation.

Because the vast majority of California mining operations will not have to comply with SB 610's water supply assessment requirement, this chapter does not address the issue further other than by referring readers to a California Department of Water Resources publication entitled, "*Guidebook for Implementation of Senate Bill 610 and Senate Bill 221 of 2001*" (www.owue.water.ca.gov/Guidebook.pdf); *see also California Oak Found. v. City of Santa Clarita* (2005) 133 Cal. App. 4th 1219, 1239–40 (invalidating an EIR for a proposed industrial development because, among other reasons, substantial evidence did not exist to support the lead agency's conclusion that sufficient water sources existed for the project).

Mining projects that may result in water discharges must consider the operations' affects on water quality.

Water quality. In addition to water supply issues, mining project CEQA documents must also address potential water quality impacts associated with the project. If not recycled, the water used by mining operations must be discharged after use. Mining projects located near or hydrologically connected to streams, rivers, or other drainages flowing to streams and rivers must analyze whether the operations will affect those water bodies. Mining projects can affect water quality if they discharge process water containing sediment, chemical constituents, or highly mineralized water into local water bodies. (Chapter 11 provides an overview of the water quality regulations that may apply to mining projects.) For purposes of complying with CEQA, environmental documents prepared for mining projects must analyze whether the proposed project will affect water quality in the vicinity of the project. Discharges of mining process water may be considered a significant impact to water quality.

Mining operations' discharge of process water into rivers and streams may also cause significant impacts to biological resources. Because terrestrial and aquatic species rely on rivers and streams for hydration and habitat, discharge of untreated mining process water can directly or indirectly affect flora and fauna. In extreme cases, toxic chemicals can leach from mining operations and poison plants and animals that come into contact with such discharges. *See* Pub. Res. Code § 21151.7 (requiring preparation of an EIR for any open pit mining operation which is subject to the permit requirements or reclamation plan requirements of SMARA and which utilizes cyanide for producing gold or other metallic minerals). Such occurrences are rare, particularly for surface mines (as opposed to certain pit mining operations) which typically do not use large amounts of chemicals in their processes.

For surface mines, sedimentation in streams and rivers caused by discharges or mining-related erosion is the primary concern. Sedimentation can cause significant adverse impacts by filling gravel beds used by salmonids for spawning with silt. Filling in gravel beds adversely affects fish egg viability because it blocks eggs from settling into the crevices created by pebbles and gravel, thereby increasing the risk that deposited eggs will be swept downstream. Potential impacts such as this should be taken into account when the CEQA documents for the mining project are prepared.

Because nearly all mining projects add particulate matter to the air, their EIRs must address air quality impacts. The cumulative impacts of the mining project, when combined with existing air quality in the local air basin, should be carefully considered.

Air quality impacts. Surface mining is an intensive land use that literally rips and consumes earth; the soil is then often loaded onto machinery and vehicles that burn fossil fuels. These characteristics of mining projects often cause significant air quality impacts. In addition, these already significant impacts may exacerbate existing poor air quality in the region in which the proposed mining project will be built. This is because many of California's major air basins have not attained the air quality compliance levels dictated by the federal Clean Air Act as implemented by California's air quality management districts. *See* United States Environmental

Protection Agency, *Region 9 Air Programs: Particulate Matter (PM_{10}) Attainment Designations in Region 9* (Sept. 2007) (www.epa.gov/region09/air/maps/r9_pm10.html).

Air quality mitigation measures imposed by lead agencies through the EIR process and conditions of approval help lessen mining project-related air quality impacts, but often do not mitigate such impacts to below the level of significance. *See, e.g., Fairview Neighbors v. County of Ventura* (1990) 70 Cal. App. 4th 238, 244. Even where project-specific air quality impacts are mitigated below the level of significance, it is still possible that, due to the influence of other projects affecting air quality or the degraded nature of the air basin where the proposed project is located, the project could have cumulatively significant air quality impacts. *See* CEQA Guidelines section 15130 (setting forth requirement to discuss cumulative impacts in an EIR) and section 15355 (defining cumulative impacts). As explained above, when a proposed project has project-specific or cumulative air quality impacts that cannot be mitigated below the level of significance, the lead agency must adopt a Statement of Overriding Considerations and weigh the project's benefits against its environmental costs before it can approve the project.

The reality is that nearly all projects (mining or otherwise) for which an EIR has been prepared will require a Statement of Overriding Considerations to be approved. Projects often affect one or more CEQA impact areas in ways that simply cannot be feasibly mitigated below a level of significance.

The reality is that nearly all projects for which an EIR has been prepared will require a Statement of Overriding Considerations to be approved.

Other environmental impacts. While a mining project could involve many other potential impacts, a few require brief mention here. Although mining projects may cause potential impacts to water and air resources, project applicants and lead agencies should also be cognizant of potential noise, biological resources, and aesthetics impacts. Because surface mines often disturb dozens, if not hundreds of acres, through the use of heavy machinery and motorized transports, it should come as no surprise that mining projects can be noisy. Likewise, when sensitive species or their habitat is present on the mining site, those biological resources are likely to be disturbed. Finally, the very nature of surface mining requires a disturbance of the soil that may affect visual or scenic resources. All of these impacts, and any other potential impacts that the lead agency is aware of, must be taken into account during the CEQA process.

chapter 9

Nonconforming Surface Mines

While chapter 7 discussed the laws governing the city and county regulation of new mines, most surface mines operating today are fairly old. Many have been operating for several years, some even for decades. Because of their ages, these mines may have long predated zoning, making them **nonconforming** uses, which do not comply with zoning but may nonetheless continue. This chapter discusses the general principles and special rules and doctrines that govern these.

General Principles Governing Nonconforming Use

Nonconforming uses are uses that preexist, and do not conform with, zoning restrictions. Zoning ordinances allow them to continue to avoid constitutional problems that would arise from compelling immediate compliance.

By definition, a nonconforming use does not comply with zoning requirements and results from having existed—in some cases, by years—prior to restrictive land use regulation. A zoning ordinance customarily permits the use to continue because compelling immediate compliance may be unconstitutional. *Jones v. City of Los Angeles* (1930) 211 Cal. 304, 310–11. Mining and similar industrial operations, for instance, usually involve huge financial investments in plants, equipment, and other facilities. Subjecting such uses to zoning after the fact might destroy those investments, resulting in unconstitutional takings of property without just compensation. *Hansen Bros. Enters., Inc. v. Bd. of Supervisors* (1996) 12 Cal. 4th 533, 551–52. A provision exempting nonconforming uses thus "is ordinarily included in zoning because of the hardship and doubtful constitutionality of compelling [their] immediate discontinuance." *County of San Diego v. McClurken* (1951), 37 Cal. 2d 683, 686.

Courts follow a strict policy against the expansion of nonconforming uses, prohibiting them from enlarging, expanding into new areas, adding new activities, and operating more intensively.

Although a nonconforming use may continue, its right to do so is limited. Courts follow a "strict policy" against expansion. *Id.* 687. Under this policy, "the continued nonconforming use must be similar to the use existing at the time the zoning ordinance became effective...." *Edmonds v. County of Los Angeles* (1953) 40 Cal. 2d

642, 651. The use accordingly may not be enlarged or extended into other areas or properties. *Rehfeld v. City and County of San Francisco* (1933) 218 Cal. 83, 85; *City of Yuba v. Cherniavsky* (1931) 117 Cal. App. 568, 573. It also may not be changed to include activities not underway when the use became nonconforming. *Paramount Rock Co., Inc. v. County of San Diego* (1960) 180 Cal. App. 2d 217, 230. Nor may the use be operated in such a way as to make it more intensive than it was when it became nonconforming. *County of Orange v. Goldring* (1953) 121 Cal. App. 2d 442, 447–48.

Nonconforming uses must operate continuously after becoming nonconforming. If they cease operating for a prescribed period, they are considered abandoned, and must then immediately comply with zoning.

To maintain its status, a nonconforming use must also operate continuously after inception. Should it cease for a prescribed period—for 180 days, a year, or two years, depending on the governing ordinance—the nonconforming use is considered abandoned. In that situation, the use must come into compliance with zoning requirements. *Hansen Bros.*, 12 Cal. 4th 569.

Every nonconforming use, mining as well as non-mining, is covered by the anti-expansion and abandonment principles discussed below. However, the uniqueness of mining has led to several refinements. This chapter explores why nonconforming uses are prevalent and likely to persist, making the rules and holdings ones that are likely to garner significant attention in the coming years.

Endurance of Nonconforming Mines

When zoning ordinances were adopted, many thought that nonconforming uses would eventually disappear. That belief has been proven incorrect.

When zoning ordinances were first adopted, many thought that nonconforming uses would disappear over time, eventually causing every use to come into compliance. Early cases in particular predicted their "gradual elimination." *See, e.g., Rehfeld v. City and County of San Francisco* (1933) 218 Cal. 83, 84. Most cases, in fact, include a statement that reads, "[z]oning legislation looks to the future in regulating district development *and the eventual liquidation of nonconforming uses within a prescribed period* commensurate with the investment involved." *Livingston Rock & Gravel Co. v. County of Los Angeles* (1954) 43 Cal. 2d 121, 127 (emphasis added). The underlying belief is that existing nonconforming uses "would be of little consequence and... would eventually disappear." *City of Los Angeles v. Gage* (1954) 127 Cal. App. 2d 442, 454.

Mining operations have often benefited from their nonconforming status. Because new, competing operations have not started up around them, they have gained a favorable economic position.

That has proven incorrect. Rather than fade into obsolescence, many nonconforming uses flourished. This is chiefly because the zoning ordinances that made them nonconforming also prohibited new competing uses from springing up nearby—giving nonconforming uses significant business advantages and, in some cases, monopolies. *Id.* 454 n.2.

Mining has been no stranger to this effect. A mine operating as a nonconforming use is often only one of a few mines in a particular market area. Indeed, the almost Herculean task of securing all the required entitlements for a new mine, discussed in chapter 2, is an expensive, years-long process. Because of intense public opposition, a new proposal may not be approved. As a result, market barriers to entry of new mines give nonconforming mines favored economic positions.

Under some circumstances, the longevity of nonconforming mining uses results from a softening of the anti-expansion principle. Although a nonconforming use may not expand by changing to a different use, courts have nonetheless allowed for new technologies and modernization of facilities and equipment. *See, e.g., Endara v. City of Culver City* (1956) 140 Cal. App. 2d 33, 38. Courts have also allowed a nonconforming use to expand production levels to meet increased market demand. *See Hansen Bros. Enters., Inc. v. Bd. of Supervisors* (1996) 12 Cal. 4th 533, 573.

Nonconforming mining operations have benefited from the softening of the rule that nonconforming uses not be expanded. They have been allowed to modernize their plants and equipment and to increase their production levels to meet market demand.

Surface mining has benefited from these decisions. Had they been left with the plants and equipment of 30 to 50 years ago, when first zoned, most nonconforming mines might not have been able to operate efficiently. Similarly, had they been confined to the production levels existing at the time they became nonconforming, many mines would not have been able to satisfy the needs of the markets they served.

In summary, rather than wither, nonconforming mines have endured, and this has created tensions with surrounding uses. As shortages become more acute in the coming years, the tensions with neighborhoods will only intensify. Granting too few permits will place greater demands on existing mines to meet the state's ever-increasing need for construction materials. At the same time, the encroachment of urban uses on mining described in chapter 2 will expose more residential and sensitive uses to mining's externalities. As these tensions play out, the rules and holdings discussed below are likely to be increasingly visited.

Determining the Existence and Scope of Nonconforming Mining Rights

Six questions must be explored when determining the existence and scope of nonconforming mining rights. After these questions are summarized here, the rules and holdings applicable to each will be discussed in the sections below.

First, when did the use become nonconforming? The answer to this question allows one to judge whether the owner's activity was substantial enough to vest the right to continue mining. If sufficient, the level of mining activity existing on the date is the baseline for establishing the permissible scope of the use. The baseline, in turn, allows one to determine whether the mine owner has expanded the scope through subsequent activity.

The first question to address in determining the existence and scope of nonconforming mining activity is when the use became nonconforming.

Second, had mining commenced by the time the restrictive ordinance was adopted? Often this question is easily answered: substantial mining may have been underway when the local zoning regulation was adopted. But this is not always the case. Both in the past and now, opening a new mine requires considerable preparation. Even if excavation had not begun when the ordinance was adopted, the mine owner may have incurred substantial expense to explore the site and obtain equipment, plants, and other necessary facilities. Thus, even when excavation may not have begun at the time a zoning ordinance was adopted, the courts may consider the operation to have commenced such that its rights as a nonconforming use are vested.

The second question for determining the existence and scope of a nonconforming use is whether mining had commenced by the time the applicable zoning ordinance was adopted.

The third question for determining the existence and scope of a nonconforming mining use is what are the activities that comprise the use that became nonconforming.

Third, what use became nonconforming? The scope of a nonconforming use only includes activities that were part of the operation when the use became nonconforming. Identifying the specific activities underway when zoning was first adopted determines the use that may continue. Some operations may have involved only extraction and sale of sand and gravel when the use became nonconforming, while others may have employed processes for crushing, processing, and washing, or the production of concrete or asphalt. If part of the use when the mine became nonconforming, the activities may continue. New activities—such as the addition of crushing or asphalt production to a mine not previously engaged in those activities—cannot be initiated after the use became nonconforming.

The fourth question concerning the existence and scope of a nonconforming mining use is what area the use may occupy.

Fourth, what area may the use occupy? Here, the **diminishing asset doctrine** applies. Recognizing that mineral extraction, unlike any other use, must move to a new area when a resource is depleted, the doctrine is sometimes viewed as an exception to the anti-expansion principle. It allows an owner to mine all areas proposed to be mined at the time the operation became nonconforming, even if that means the unmined areas will lie in reserve for many years after the restrictive ordinance was adopted.

The fifth question relating to the existence and scope of a nonconforming mining use is how, if at all, the use has changed over time.

Fifth, how may the use change over time? Often many years old, a nonconforming mine may have been modernized, adding to plant and equipment as mining technology improved. Others may have experienced significant business growth, leading to production levels exceeding what existed when the mines became nonconforming. Several court cases have considered the extent to which changes in mining operations are permissible, and some have relaxed the anti-expansion principle, allowing a mine to modernize and expand production to meet increased market demand. But the cases have also held that intensification is improper when the nature of the use changes or affects its surroundings in a new and adverse way.

The sixth and final question for determining the existence and scope of a nonconforming mining use is whether the operator has abandoned its right to continue the use.

Sixth, has the owner abandoned the right to continue the use? Like any other nonconforming use, a mining use may not cease for a significant period without terminating its exemption from zoning requirements. That said, surface mining is cyclical, an industry that rises and falls in lock-step with the economy. During any significant downturn, a mine may close for a prolonged period, and courts consider economic conditions when evaluating nonconformance. Courts have generally held that under these circumstances a mine owner does not abandon the right to continue the mine as a nonconforming use.

When a Mining Use Becomes Nonconforming

Determining the date on which the mining operation became nonconforming is critical because it is the date for which the activity concerning the existence and scope of the use is judged.

To determine the existence and scope of a nonconforming use, the first step is to identify when the mine became nonconforming. This is critical because mining must have commenced by that date for a mine owner's rights to vest. Whether it is an expanded use is evaluated by comparing future activity to activity that existed on the date.

A use can become nonconforming in two ways. A zoning ordinance can prohibit the use—for instance, by placing an existing mine in a zoning district that does not permit mining under any circumstances. *See, e.g., Rehfeld v. City and County of San Francisco* (1933) 218 Cal. 83, 84. Or the ordinance can make it a **conditional use**—one that permits mining, but only after securing a discretionary permit. *See, e.g., Stokes v. Bd. of Permit Appeals* (1997) 52 Cal. App. 4th 1348, 1351. In the latter instance, while not prohibited, mining is an unlawful use if engaged in without obtaining a permit and abiding by its terms.

A mining use can become nonconforming in two ways. A zoning ordinance can prohibit mining altogether or it can make mining a conditional use, one that is unlawful unless conducted with a permit.

By the 1940s or 1950s, most local agencies had adopted zoning ordinances that gave rise to nonconforming mining uses. In many cases they only applied to certain areas, leaving the mining unregulated, particularly if the operations occurred in isolated locations. Other ordinances placed mine property in an **unclassified** or catch-all district, permitting as-of-right any use "not otherwise prohibited by law." In those instances, mining would have been a conforming use until a subsequent regulation was adopted that either prohibited mining altogether or made it conditional.

Regardless of local requirements, however, mining operations not regulated as of 1976 became nonconforming with the passage that year of the State Surface Mining and Reclamation Act (SMARA). Although intended to regulate reclamation, and not mining's land use entitlements, the Act nonetheless provides: "[N]o person shall conduct [a] surface mining operation unless a permit is obtained from... the lead agency...." Pub. Res. Code § 2770(a). SMARA defines **permit** as "any *authorization from, or approval by,* a lead agency, the absence of which would preclude surface mining operations." *Id.* § 2732.5 (emphasis added). Taken together, these provisions require that a mining operator formally obtain a permit from the lead agency before commencing the activity. As explained in chapter 7, these SMARA provisions may have made mining in every jurisdiction at least a conditional use, as of 1976, even if its local ordinance did not reflect that fact.

Because of the SMARA permit requirement, mining uses that did not become nonconforming by 1976 likely became nonconforming no later than that year.

SMARA = Surface Mining and Reclamation Act

Whatever the actual date of nonconformity, an owner must demonstrate that mining activity commenced by that date.

Activities Required to Vest Rights in a Nonconforming Surface Mine

The next step in determining if a surface mining operation may continue as a nonconforming use is to evaluate whether the operation commenced by the date zoning restrictions became effective. Often, this issue causes no dispute. If a sizable area had been opened by the date of the change, one could not plausibly contend that an owner's rights did not vest. Even if minimal when zoning went into effect, the existence of excavation makes it impossible to contend that the operation had not commenced. However, a more difficult question arises when much of the preparatory work was completed, but excavation had not begun. In such a situation, one would be tempted to posit that mining did not commence because there was not then a "hole in the ground."

A mining operator acquires a vested right when it has commenced surface mining operations or has incurred substantial liabilities in preparation for mining by the date the operation becomes nonconforming.

Yet the standard for commencement is not so demanding. An owner's rights could have vested from preparatory activities that required a substantial investment before the restrictive ordinance was adopted. SMARA recognizes this principle, providing that a person is deemed to have a vested right to continuing a nonconforming use when "prior to January 1, 1976, he or she has, in good faith... diligently commenced surface mining operations *and incurred substantial liabilities for work and materials necessary therefor.*" Pub. Res. Code § 2776 (emphasis added). SMARA's definition of **surface mining** makes clear, moreover, that a mining operation includes "[p]rospecting and exploratory activities" as well as excavation. *Id.* § 2735(c). Thus, activities short of excavation could vest nonconforming mining rights as long as they imposed "substantial liabilities" on the mine owner.

If mineral production had not commenced by the time mining use became nonconforming, activities relevant to whether the operation has acquired a vested right include prospecting and exploration, purchase or leasing of the mine site, hiring of employees, installation of plants and equipment, construction of a road system, land clearing, and overburden removal.

Defining "substantial liabilities" is a question of fact unique to each operation. Nonetheless, opening a mine is an activity that—today as well as years earlier—requires substantial time and a huge capital investment. Before the first ton of material is excavated on the property, an owner has to prospect and explore potential sites, making arrangements for purchase or lease. Then a workforce has to be secured as well as plants, equipment, and facilities necessary to process, store, and load the excavated material. The owner must then construct a road system to connect the operation to areas for processing, storage, and staging, clear vegetation, and strip the overburden to expose the ore.

Overall, these activities require considerable time and planning, along with significant expense. An owner would not be able to recover any of this if a zoning ordinance adopted just prior to initial excavation precluded a mining operation. Because such a result would be unduly harsh, an owner's rights in a nonconforming use can vest even though excavation had not begun at the time of a zoning change. While the specific facts for each project may differ, the sufficiency of the investment to justify vesting an owner's rights must stand on its own. As indicated in Figure 9-1, a few preparatory acts are not likely to suffice. *See* 59 Op. Cal. Att'y Gen. (1976) 641, 653 (stating that "a nominal amount of 'prospecting and exploratory activities' on particular land does not establish a vested right"). An owner would most likely have to be fairly well along in the process for the rights to vest.

A few preparatory acts would probably not be enough to establish a vested right.

Activities That May Be Part of a Use That Becomes Nonconforming

If mining had commenced by the date a restrictive ordinance was adopted, an operator may only perform activities within the scope of the nonconforming use, subject to the limited exceptions described below. *Paramount Rock Co., Inc. v. County of San Diego* (1960) 180 Cal. App. 2d 217, 228. Activities identified at this stage of the analysis constitute the baseline against which comparisons of future activity are made

Figure 9-1

Establishing a Vested Right

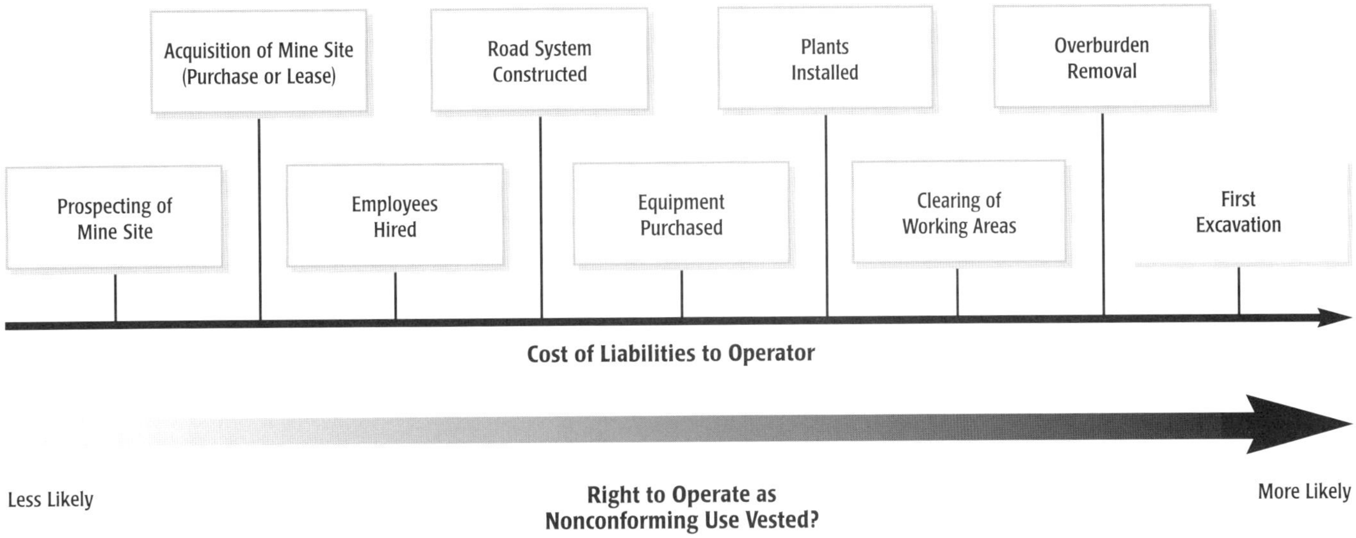

when determining whether a use has expanded impermissibly. *See Edmonds v. County of Los Angeles* (1953) 40 Cal. 2d 642, 651 (nonconforming use must remain similar to use existing at the time it became nonconforming).

Determining the scope of permissible activities is not difficult: a nonconforming use includes all activities that were "integral parts" of the mining operation at the time a restrictive zoning ordinance was adopted. *Hansen Bros. Enters., Inc. v. Bd. of Supervisors* (1996) 12 Cal. 4th 533, 542. Applying this standard, one is tempted to focus on the obvious—extraction of minerals. However, the California Supreme Court has observed that a nonconforming operation involves more than the mining of minerals and materials; it includes all "uses normally incidental and auxiliary to the nonconforming use." *Id.* 565. Activities properly considered part of a mining use include "crushing, combining and storing the mined materials" as well as "selling or trucking the aggregate from the property." *Id.* 542. Likewise, the use may involve importation of raw materials for blending and processing with material extracted onsite. *See Mass. Broken Stone Co. v. Town of Weston* (Mass. 1964) 195 N.E. 2d 522, 526. Office buildings needed for mine administration may also be considered part of a nonconforming mining use. *See Hilltown Township v. Horn* (Pa. Commw. Ct. 1974) 320 A. 2d 153, 156.

Because a nonconforming use may not add any activity beyond what was integral to mining at the time the use became nonconforming, one should take care to identify all activity underway at the time the zoning was changed. Even if carried on for many years, an owner can be ordered to cease an activity if a lead agency discovers that it was not originally part of the use.

A nonconforming mining use includes all activities that were integral parts of the operation at the time it became nonconforming. These activities can include crushing, processing, storage, trucking, importation of raw materials, and the use of administrative offices.

PRACTICE TIP

When documenting a nonconforming mining use, a good place to start is the California Geological Survey Library in Sacramento. The library has historical volumes, articles, and reports with coverage of mining operations throughout the state as far back as the late 19th century. Aerial photography may also be a good way to document a nonconforming use. In many cases, preparatory acts will be evident from an aerial photo of the mine site.

Area a Nonconforming Use May Occupy: the Diminishing Asset Doctrine

✍ **PRACTICE TIP**

Because a nonconforming use is confined to any activity carried on when it became nonconforming, every activity then existing should be identified. Aerial photography may be the best way to document crushing and processing, as well as asphalt and cement production, since these and other activities are visible from the air.

Visiting a working mine site, one cannot help but be impressed by the huge amount of land removed. Looking across the property, a sand and gravel pit may extend for several acres, while a hard rock operation may descend hundreds of feet below, or ascend upward in rings of benches in a quarry pit a few thousand feet wide.

But the conditions observed did not suddenly become that way. The land disturbance likely took years, perhaps even decades, to take shape. Before mining, the land may have been densely vegetated, or hilly, possibly even bucolic. If shown a photograph of the land in its original state, a visitor may not even recognize that it is the same property.

The changes these properties undergo are a testament to the uniqueness of mining. Unlike other uses that occur on fixed territory, a mining land use must expand into new areas. Mineral deposits are often spread throughout the property, and, over a mine's lifetime, the operation will progress from one area to the next, moving on as each deposit is exhausted. After many years, enormous amounts of surface land will have been disturbed.

But the pace of this process is slow. Many mine sites are large, made up of hundreds of acres, and reserves may be sufficient for an operation to continue 50 to 100 years or more. During the early stages of its lifetime, very little property may be disturbed. What happens, then, if a zoning ordinance is adopted at such a stage, when excavation occurs on only a small portion of the overall property to be mined?

Because mining by its nature cannot take place in fixed areas, the diminishing asset doctrine exists to prevent the harshness of the literal application of the rule prohibiting nonconforming uses from expanding.

Applying the anti-expansion principle too literally in this instance would prohibit an owner from mining in new areas on the property, effectively forcing an end to the operation once existing sites were mined out. Yet that result would be harsh, since an owner may reasonably have expected to continue many more years of mining given the size of the property and extent of its reserves. Prevented from fulfilling such an expectation, an owner might claim that the property had been taken without just compensation, which would likely make a compelling case.

To address this situation, the courts have developed what is known as the diminishing asset doctrine. Formulated over many years, this doctrine allows an owner to work all the land intended for mining at the time the use became nonconforming.

California's Recognition of the Diminishing Asset Doctrine

Surface mining, in contrast to a static land use operating within fixed parameters, is progressive. New areas are consumed as deposits in previously mined areas are depleted. As one court observed, "the very nature and use of an extractive business contemplates the continuance of such use of the entire parcel of land as a whole, without limitation or restriction to the immediate area excavated at the time the ordinance was passed." *McCaslin v. City of Monterey Park* (1958) 163 Cal. App. 2d

339, 349. Given this reality, a literal application of the prohibition against expansion of nonconforming uses would force an owner to stop mining almost as soon as the operation became nonconforming. Such a result would "practically amount to a taking of the property since it [would] den[y] the owner the right to engage in the only business for which the land is fitted." *Id.* (citing *Morton v. Super. Ct.* (1954) 124 Cal. App. 2d 577, 582).

If literally applied, the anti-expansion principle would force a mine operator to stop mining almost as soon as its operation became nonconforming.

The diminishing asset doctrine avoids the harshness and possible unconstitutionality of such a result. The California Supreme Court recognized its applicability in a seminal case, *Hansen Bros. Enters., Inc. v. Bd. of Supervisors* (1996) 12 Cal. 4th 533, summarizing the issue as follows:

The California Supreme Court recognized the diminishing asset doctrine in a seminal case, Hansen Bros. Enters., Inc. v. Bd. of Supervisors.

> Unlike other nonconforming uses of property which operate within an existing structure or boundary, mining uses anticipate extension of mining into areas of the property that were not being exploited at the time a zoning change caused the use to be nonconforming. The question thus arises whether this extension is a prohibited expansion of a nonconforming use into another area of the property. In those jurisdictions which have considered the question, the answer is a qualified "no" under the "diminishing asset" doctrine, an exception to the rule banning expansion of a nonconforming use that is specific to mining enterprises.
>
> *Id.* 553

In adopting the doctrine, the Court relied heavily and quoted liberally from several out-of-state cases. It held that mining may expand into unmined areas after adoption of a zoning ordinance only "[w]hen there is objective evidence of the owner's intent to expand a mining operation, and that intent existed at the time of the zoning change...." *Id.* Indicating on several occasions its adoption of an objective standard, the Court made clear that "[t]he mere intention or hope on the part of the landowner to extend the use over the entire tract is insufficient; the intent must be objectively manifested by the present operation." *Id.* 557 (citations omitted).

The diminishing asset doctrine allows nonconforming mining operators to expand to areas where objective evidence indicates they intended to mine when they became nonconforming.

While recognizing the diminishing asset doctrine and outlining its basic requirements, *Hansen Brothers* did not spell out specific activities and circumstances that would objectively manifest an intent to mine. The Court instead remanded the case for application of the key factual issues. As a consequence, *Hansen Brothers* does not offer much guidance as to how to apply the diminishing asset doctrine. Nonetheless, given the Supreme Court's reliance on several out-of-state cases, it is reasonable to assume that those cases would be persuasive in identifying common indicators reflecting an intent to mine under the doctrine.

Factors Applicable to Diminishing Asset Doctrine Determinations

Even before *Hansen Brothers*, many cases had considered the diminishing asset doctrine. Indeed, a California appellate court recognized the doctrine nearly 40 years earlier. *See McCaslin v. City of Monterey Park* (1958) 163 Cal. App. 2d 339, 349.

Other diminishing asset cases were decided around that time, with the earliest to recognize the doctrine decided in 1932. *See Lamb v. A. D. McKee* (N.J. 1932) 160 A. 563, 564. When the California Supreme Court considered its applicability in 1996, the doctrine had already attained a well-established pedigree.

There is no formula or bright line test for applying the diminishing asset doctrine. However, one can distill some common indicators from the cases.

Despite the large number of relevant cases, courts have not identified a formula or bright line test for applying the doctrine. However, a review of the cases points to some common indicators for determining when particular uses satisfy the doctrine's requirements and when they do not.

Length of mining operations before adoption of restrictive zoning ordinance. One important indicator of an owner's intent can be gleaned from the amount of elapsed time between commencement of mining and passage of the ordinance making the use nonconforming. Generally, the longer property has been mined before adoption of the ordinance, the more likely an owner's intent to mine the property in its entirety. In a New York case, for instance, the court noted that a mine owner had "engage[d] in substantial quarrying activities on a distinct parcel of land over a long period of time and *these activities clearly manifest an intent to appropriate the entire parcel to the particular business of quarrying. . . .*" *Syracuse Aggregate Corp. v. Weise* (N.Y. 1980) 414 N.E. 2d 651, 655 (emphasis added).

The amount of time between the commencement of mining and the passage of the ordinance making the mining operation nonconforming is an indicator of the operator's intent.

By contrast, in other cases a shorter period between initial activity and passage of the restrictive ordinance has led to the opposite conclusion. This has particularly been so when another indicator of intent—the size of the excavation in relation to the size of the property—is present. If only a small percentage has been disturbed in the brief period preceding adoption of the ordinance, courts are reluctant to allow an owner to mine a vastly greater area. *See, e.g., Skenesborough Stone, Inc. v. Vill. of Whitehall* (N.Y. App. Div. 2000) 708 N.Y.S. 2d 171, 173–174 (minimal preparatory activity and limited excavation shortly before site became nonconforming did not evidence intent to mine entire 400-acre property).

Operators have had difficulty proving an intent to mine large tracts of reserve land when the size of their excavation was small at the time of the zoning ordinance's adoption.

Size of excavated area compared to the size of total mine property. A related indicator is the size of the excavation as a portion of the overall property. Owners claiming a right to mine huge tracts of reserve lands have had difficulty presenting objective evidence that mining those lands had always been their intention. In a New York case, the court rejected the contention that an operator could mine all 400 acres owned. *Skenesborough Stone, Inc. v. Vill. of Whitehall* (N.Y. App. Div. 2000) 708 N.Y.S. 2d 171, 174. The court noted that a mere eight to ten tractor loads of material had been removed from easily accessible areas of the property before adoption of the ordinance. *Id.* 173–74. That activity, the court concluded, "failed to establish substantial mining activities over any significant period of time manifesting an intent to appropriate the entire 400-acre site to the particular business of mining." *Id.* 174.

Faced with limited excavation on large property, a local agency might be justified in limiting future expansion to less than the entire property. In *Stephan & Sons,*

Inc. v. Municipality of Anchorage Zoning Board of Examiners & Appeals, the court upheld a local board decision to restrict expansion to only 13 of 55 acres. The court noted that gravel operations had occurred on only two to five acres in the four years between commencement of the operations and adoption of the ordinance. (Alaska 1984) 685 P. 2d 98, 102. "Given these facts," the court opined, "we believe the Board would have been justified in restricting Stephan's nonconforming use to even less than the thirteen acres which it approved." *Id.*

Use of reserve property as support for mining operations. Mining requires that excavation occur in certain areas at any one time, with new areas exploited only when deposits in older areas yield. This attribute is unique to mining, since large tracts may lie in reserve for many years before being disturbed. While difficult to establish an intent to mine substantial reserved land in its entirety, such a showing is not impossible. The cases reveal two indicators that have allowed an owner to demonstrate an intent to mine the entire property: first, as explained here, much of the unexcavated area was used as support for the mining operations, and second, as explained next, unexcavated areas had been prepared for mining.

Mining involves more than excavation. A mine usually has areas devoted to plants and equipment for crushing, processing, and washing operations along with offices for administration and areas for storage and stockpiling. Facilities for loading trucks and truck staging are common along with a system of onsite roads to allow heavy equipment access to various locations on the property. When all areas are included, the size of the operation is much larger than an excavation site. Courts have therefore allowed excavation to extend into areas connected in some way with the overall operation even when not mined at the time of a zoning change. *See County of DuPage v. Elmhurst-Chicago Stone Co.* (Ill. 1960) 165 N.E. 2d 310, 312 (future intended excavation area was connected to mining areas existing when use became nonconforming because it was used for stockpiling and had switch tracks connecting it with other mine areas); *Bainter v. Vill. of Algonquin* (Ill. App. Ct. 1996) 675 N.E. 2d 120, 123 (tract was part of nonconforming mining use because it had been connected by underground tunnel to tract on which mining was occurring at time of zoning change); *see also Gibbons & Reed Co. v. N. Salt Lake City* (Utah 1967) 431 P. 2d 559, 564–65.

Where reserve land is connected in some way to a mining operation, courts have often concluded that the operator intended to mine those areas.

Preparation of reserve areas for mining. Even if mine property had not been excavated when the use became nonconforming, courts have found that certain preparatory activities reflect an intent to mine those areas. The construction of a road system that would allow heavy machinery access to reserve areas in the future has been one important activity. *See Gibbons & Reed Co. v. N. Salt Lake City* (Utah 1967) 431 P. 2d 559, 565; *Syracuse Aggregates v. Weise* (N.Y. 1980) 414 N.E. 2d 651, 655. Another preparatory activity is the clearing or timbering of reserved lands. *Town of W. Greenwich v. A. Cardi Realty Assocs.* (R.I. 2001) 786 A. 2d 354, 364. The construction of a berm enclosing reserve areas likewise has reflected an intent to mine reserve areas in the future. *See Bainter v. Vill. of Algonquin* (Ill. App. Ct. 1996) 675 N.E. 2d 120, 123.

The preparation of reserve areas for mining—through the connection of those areas to a mine site's road system or activities such as clearing—is an additional indicator that the operator intended to mine those areas.

Where operators owned separate parcels at the time their uses became nonconforming, courts have often found that the reserve parcels were not part of their nonconforming uses.

Existence of separate parcels and artificial barriers. Mining property is usually extensive, composed of numerous parcels where extraction occurs while others lie in reserve. For many owners this can be problematic. Except when tangibly connected to parcels where extraction is taking place or showing signs of preparation, courts usually do not consider reserve parcels part of the nonconforming use. *See Dolomite Prods. Co., Inc. v. Kipers* (N.Y. App. Div. 1965) 260 N.Y.S. 2d 918, 921 (observing "[i]t is not consonant with progressive or contemporary planning to permit one to purchase a large parcel of real property, work thirty-five acres of it and do nothing for 40 years within [the other parcel] but, nevertheless, have the right some time in the distant future to make a non-conforming use of it in violation of an ordinance prohibiting it and to the great detriment of adjacent homeowners").

Where roads or highways have separated parcels, courts have usually found that reserve parcels were not part of the operators' nonconforming uses.

Artificial barriers present the same dilemma. When roads separate the active and reserve areas of the property, courts usually consider reserve areas distinct and not part of the nonconforming use. *See Fred McDowell, Inc. v. Bd. of Adjust. of Township of Wall* (N.J. Super. Ct. App. Div. 2000) 757 A. 2d, 822, 827 (construction of freeway separated mine lot and reserve lot, making access to the reserve lot from the mine lot impractical).

Courts have often concluded that the use of reserve areas for non-mining uses is inconsistent with an intent to mine.

Use of reserve areas for non-mining uses. The use of reserve areas for a non-mining purpose may indicate that an owner does not intend to mine all of the property—for instance, when leasing property to a farmer. *See R.K. Kibblehouse Quarries v. Marlborough Township Zoning Hrg. Bd.* (Pa. Commw. 1993) 630 A. 2d 937, 944. That might not be the case when reserve areas benefit from a non-mining use. In a Rhode Island case, the court found the use of reserve property for grazing consistent with an intent to mine, since grazing kept the area free of vegetation that would otherwise have to be removed to facilitate mining. *See Town of W. Greenwich v. A. Cardi Realty Assocs.* (R.I. 2001) 786 A. 2d 354, 361.

Expansion of Nonconforming Mining Uses

Many mining uses are old, some having begun as long as a century ago. During this period, technology has improved dramatically. What once required a few hours or even a day can now be produced in an hour. New highways, roads, and developments have likely sprung up nearby, thanks in large part to the minerals and construction materials the mine made available. And because of population growth, what probably started out as a small mine in an isolated area is now a full-scale operation producing a million or more tons each year.

Had the prohibition against expansion of nonconforming uses been applied literally, these changes would not have been permissible. Strict adherence to the anti-expansion principle would have mandated that conditions be frozen when the mine became nonconforming—and the mine would only have been able to employ plants, equipment, and levels of production that existed at the time.

The law, however, is not inflexible. Within reasonable limits, a nonconforming mine may modernize facilities and increase production in response to increased demand due to population growth. These exceptions, like the diminishing asset doctrine, recognize the realities of surface mining and seek to strike a balance between an operator's right to continue and the principle that nonconforming uses not expand. But the exceptions do not give an operator *carte blanche* authority to overhaul plants and equipment or dramatically increase production. As discussed below, the exceptions are limited.

Modernization

The law recognizes that mining uses will need to modernize and increase production over time to meet demand.

Courts have been flexible in allowing a nonconforming mine to change plants and equipment as technology evolved. If the operation originally involved removal by hand, an owner was not prohibited from later removing the material by machine. *See Town of Wayland v. Lee* (Mass. 1950) 91 N.E. 2d 835, 839. However, courts have limited the extent of permitted modernization. While enabling a mine to do stay abreast of the times, modernization can also enable a mine to expand its product line significantly or detrimentally affect its surroundings. In those situations, courts have refused to relax the principle of anti-expansion.

Courts allow nonconforming operators to modernize, but not such that they can significantly expand production.

Courts have allowed mining operators to modernize when their new plants and facilities were improvements on existing ones.

Determining when modernization crosses acceptable limits is not an easy task. Courts have recognized that "the adoption of more modern instrumentalities, suitable in carrying on a business should not, in legal contemplation, work a change of use." *Endara v. City of Culver City* (1956) 140 Cal. App. 2d 33, 38. Beyond that and similar statements, no clear standard dictates when a change has gone too far. Nonetheless, a common characteristic in cases in which courts have allowed modernization is that the new plants or equipment were improvements of facilities that already existed. *See App. of Miller* (Pa. Commw. Ct. 1971) 281 A. 2d 364, 366–67 (addition of asphalt plant was permissible because the ready-mix concrete plant had already existed on site and the addition was necessary to produce material meeting new state road specifications); *Maloy v. Town Bd.* (N.Y. App. Div. 1983) 461 N.Y.S. 2d 529, 530–31 (replacement of screen gravel separator with mobile rock crusher was permissible). This is in contrast to situations where a mine owner added plant or equipment not previously employed. *See Paramount Rock Co., Inc. v. County of San Diego* (1960) 180 Cal. App. 2d 217, 230 (addition of rock crusher prohibited because ready-mix concrete business did not involve crushing when it became nonconforming); *Prior Lake Aggregates, Inc. v. City of Savage* (Minn. Ct. App. 1984) 349 N.W. 2d 575, 579 (addition of asphalt plant not permitted because use involved only extraction and sale of sand and gravel extraction at the time use became nonconforming). In the latter instances, the owner was able to create products not previously manufactured by installing a concrete plant, even though the use had involved only sand and gravel production when it became nonconforming. *See Conn. Sand & Stone Corp. v. Zoning Bd. of App.* (Conn. 1963) 190 A. 2d 594, 597; *see also Mass. Broken Stone Co.*

v. Town of Weston (Mass. 1964) 195 N.E. 2d 522, 526 (operation could not import material for sale onsite; at time it became nonconforming, imported material was sold only after being mixed with products extracted on site).

Courts have held that changes to a nonconforming mining operation are not appropriate if they allow an operation to increase production in a way that adversely affects its surroundings.

Another indication that modernization has gone too far is when new facilities create a "substantially different and adverse impact on the neighborhood." *Town of Wolfeboro v. Smith* (N.H. 1989) 556 A. 2d 755, 759; *see also County of Orange v. Goldring* (1953) 121 Cal. App. 2d 442, 448. New facilities that improve older ones are not permitted when the increased capacity has significant adverse effects on the operation's surroundings. In one case, a court prohibited replacement of a screen loader with a massive plant because the plant's enlarged capacity would have substantially increased the number of truck trips stemming from the nonconforming mine. *See Frank Casilio & Sons v. Zoning Hrg. Bd.* (Pa. Commw. Ct. 1976) 364 A. 2d 969, 970.

Increases in Production Volume

Population is not static, particularly in California, where nearly every region has grown substantially over the past decades. It should not be surprising, then, that most mines have also seen production levels grow. And with the looming shortages described in chapter 2, mines are likely to witness even more demand, raising a question about the extent of increase that is permissible.

Nonconforming mining uses may expand production so long as the expansions are natural and gradual.

The Supreme Court has partially answered that question in *Hansen Brothers Enters., Inc. v. Bd. of Supervisors* (1996) 12 Cal. 4th 533. The Court recognized the right to a "gradual and natural increase in a lawful, nonconforming use of a property, including quarry property." *Id.* 573. The Court elaborated:

> By way of example, we assume that a grocery store operating as a lawful, nonconforming use in an area of increasing population would not be restricted to the same number of customers and volume of business conducted when the zoning ordinance was enacted. Neither an increase in the number of patrons or in the volume of goods sold would be considered an enlargement or intensification of the use. *And where increased population creates an increased demand for the aggregate used in road construction, an increase in production to meet that demand would not be construed as an enlargement or intensification of the use.*
>
> *Id.* (emphasis added)

Beyond recognizing a general right to expand, *Hansen Brothers* does not spell out the extent to which production may increase. Whether a two-, three-, or four-fold increase is permissible is not clear, and most likely depends on how a proposed expansion would affect a mine's surroundings. In an Oregon case, for instance, the court prohibited an expansion in volume because greater production increased the blasting vibration from a nonconforming mine, created excessive noise, and generated smoke and fumes. *See Bither v. Baker Rock Crushing Co.* (Or. 1968) 438 P. 2d 988, 993–94. The court explained, "[w]here, as here, the type of nonconforming use is offensive to the prevalent use in the surrounding area, *and its volume has vastly increased since zoning,* we hold the prohibition against an increase is valid." *Id.* 994 (emphasis added).

Abandonment of a Nonconforming Mine

Mining depends heavily on the economy, demand rising when the economy grows and falling when it recedes. This is particularly true for minerals and materials when new private construction slows. Likewise, the building of roads, highways, and public infrastructure wanes as government shifts priorities in the wake of a shrinking public budget. While some can weather a downturn, others are particularly hard hit when a recession is prolonged. Some operations shut down temporarily, as occurred during the recession California suffered in the first half of the 1990s. When that happens, the question arises: By temporarily closing down because of a bad economy, does a nonconforming mine abandon the right to operate as a nonconforming use?

Nonconforming mining operations do not automatically lose the right to continue if they cease operations for certain periods.

Generally, that is not the case. While a zoning ordinance may state that cessation for some prescribed period—180 days, one year, or two years—will result in abandonment of the right to continue a nonconforming use, courts have usually ruled otherwise. As with the diminishing asset doctrine and exceptions for modernization and business growth, courts have interpreted the abandonment prohibition pragmatically in light of the realities of the mining industry.

The California Supreme Court summarized the standard for abandonment for a nonconforming use in *Hansen Brothers,* stating:

> Abandonment of a nonconforming use ordinarily depends upon a concurrence of two factors: (1) An intention to abandon; and (2) an overt act, or failure to act, which carries the implication the owner does not claim or retain any interest in the right to the nonconforming use.
>
> *Id.* 569 (citations omitted)

The critical prong of this test is the first, dealing with intention. While closing a mine for a prolonged period might constitute an overt act of cessation, the closure itself does not necessarily indicate that an owner intended abandonment of rights. Indeed, because of economic conditions, the owner may have had no choice.

Nonconforming mining operations abandon their right to continue if they cease operations and intend to abandon.

Yet, even in an economic downturn, an owner cannot maintain the right to continue a nonconforming use simply by shutting the gates and doing nothing during closure. To reflect an intent to continue, the owner must instead sell, or attempt to sell, stored and stockpiled material. *See Union Quarries, Inc. v. Bd. of County Commr's* (Kan. 1970) 478 P. 2d 181, 187 (although mine had ceased extraction, owner indicated intent to continue operations by selling from stockpiles); *Bither v. Baker Rock Crushing Co.* (Or. 1968) 438 P. 2d 988, 993 (despite cessation of mining, mine preserved right to continue by selling materials from stockpiles). The operator must also keep the plants and equipment in good order for prompt resumption of activity.

If it temporarily closes, the operator can only preserve its right to continue if it engages in activities to maintain its ability to operate.

See S. Equip. Co., Inc. v. Winstead (N.C. Ct. App. 1986) 342 S.E. 2d 524, 527. *But see County of DuPage v. K-Five Constr. Corp.* (Ill. App. Ct. 1994) 642 N.E. 2d 164, 165, 168 (dismantling plant in absence of asphalt production for ten years reflected intent to abandon such production). The operator must also continue making royalty or lease payments as it may be required to pay others. *See Union Quarries,* 478 P. 2d 187.

The longer the period of inactivity, the more likely a court will find that the operator abandoned its right to continue.

The longer the cessation of activities, however, the more likely an owner will be found to have abandoned the nonconforming use. As the California Supreme Court noted in *Hansen Bros.*, "the duration of nonuse may be a factor in determining whether the nonconforming use has been abandoned." 12 Cal. 4th 569. In two cases, nonuse for periods of seven and ten years, coupled with the absence of other preservative activity, reflected an intent to abandon nonconforming mining operations. *See Lane County v. Bessett* (Or. Ct. App. 1980) 612 P. 2d 297, 301; *Holloway Ready Mix Co. v. Monfort* (Ky. Ct. App. 1971) 474 S.W. 2d 80, 83.

Amortization of Nonconforming Mining Uses

When zoning ordinances were first adopted, the goal was to achieve uniformity within various districts. A nonconforming use was viewed as a temporary, but tolerable nuisance that would eventually become obsolete. However, as courts have since observed, the expectation that nonconforming uses would wither over time has proven incorrect. *City of Los Angeles v. Gage* (1954) 127 Cal. App. 2d 442, 454.

Perhaps nowhere is this more true than for nonconforming mines. At the time they became nonconforming, many surface mines had millions of tons of remaining reserves. Even today, the some operations' reserves may be sufficient to allow some mines to operate for decades. If zoning's goal is to separate incompatible uses, the proliferation of nonconforming surface mines may make that goal unattainable for many more years.

Because nonconforming mining uses have endured, amortization is the only scheme available to compel their termination.

This problem is not new. Fifty years ago, courts had begun to recognize that the persistence of nonconforming uses threatened the integrity of zoning. From this recognition came a solution known as **amortization**, the forced termination of nonconforming uses within prescribed periods. As one court explained, "[t]here is a growing tendency to guard against the indefinite continuance of nonconforming uses *by providing for their liquidation within a prescribed period.*" *Id.* 454 (emphasis added). By amortizing nonconforming uses, agencies could "place a time limit upon the continuance of existing nonconforming uses, commensurate with the investment involved and based on the nature of the use...." *Id.* 459. The idea behind such a scheme was to protect the owners' rights to a reasonable extent but ultimately allow the public's interest in zoning to prevail:

> Use of a reasonable amortization scheme provides an equitable means of reconciliation of the conflicting interests in satisfaction of due process requirements. As a method of eliminating existing nonconforming uses it allows the owner of the

nonconforming use, by affording an opportunity to make new plans, at least partially to offset any loss he might suffer. The loss he suffers, if any, is spread out over a period of years, and he enjoys a monopolistic position by virtue of the zoning ordinance as long as he remains. If the amortization period is reasonable the loss to the owner may be small when compared with the benefit to the public. Nonconforming uses will eventually be eliminated.

Id. 460

Amortization is therefore the only effective way to eliminate nonconforming uses. Absent that device, nonconforming uses—particularly mining uses—will continue, subject only to the prohibition against expansion.

Cities and counties must provide a reasonable time limit in amortizing nonconforming uses.

An agency employing amortization must establish a reasonable time limit for the use being amortized. To make this determination, each use must stand on its own facts. *Nat'l Adver. Co. v. County of Monterey* (1970) 1 Cal. 3d 875, 879. The relevant factors include: (1) the extent of the restriction; (2) the costs the owner will incur in finding a new suitable location; (3) the time in which the owner had notice of its nonconforming use; and (4) the extent of the nonconformity. *People v. Gates* (1974) 41 Cal. App. 3d 590, 604.

Amortization schemes for mining operations would likely have to allow for long amortization periods because of the significant investment operators must make in their operations and difficulty for them to move their operations to other locations.

Any amortization scheme applying these factors to mining must consider the realities of the industry. Mining is capital-intensive and requires substantial investment. Facilities for mining are not cheap—with some plants costing millions of dollars—and a new operation can't be started at any location. Rather, a new location must have a suitable deposit of minerals or materials. Assuming the company could obtain a permit—an exceedingly difficult task in today's climate—the new location would have to be prepared. A road system would have to be created, new plants and other support facilities built, vegetation and overburden removed, and so on—all at considerable expense to the owner. Given these realities, the amortization period for a mining use may exceed that required for another use.

The specific number of years required depend upon the facts for each case. A California court in one case upheld a 20-year amortization. *See Livingston Rock & Gravel Co. v. County of Los Angeles* (1954) 43 Cal. 2d 121, 127–28. In another, the New Hampshire Supreme Court upheld a scheme that limited the nonconforming uses from expanding in area beyond 25 percent, and from exceeding depths below the lowest grade then existing. *Flanagan v. Town of Hollis* (N.H. 1972) 293 A. 2d 328, 330. Future amortization schemes may have to provide similar parameters if an agency chooses to compel termination.

Process for Determining Right to Operate as a Nonconforming Use

As this chapter has shown, the doctrines governing nonconforming mining uses are judicially created. Balancing the harshness of requiring immediate compliance with zoning against the competing need for uniformity, courts over the past few decades have developed a number of rules concerning the existence and scope of nonconforming uses. Remarkably, though, none of the many cases decided during this period considered the process agencies must follow in addressing these issues.

In Calvert v. County of Yuba, *the court held that cities and counties must provide notice and hold a public hearing before making vested mining right determinations.*

In early 2007, however, the Third District Court of Appeal decided *Calvert v. County of Yuba* (2007) 145 Cal. App. 4th 613, which held that cities and counties must give notice and hold a public hearing before determining issues concerning nonconforming mining uses. In reaching this decision, the court struck down a county's vested rights determination made solely by its staff.

As discussed below, the *Calvert* decision seems sound as to the specific issues it resolved. As noted above, many nonconforming mining questions have flourished because of the barriers that zoning ordinances have unintentionally erected for potential competitors. With many nonconforming operations also having decades of reserves, it seems logical that the public should have input in any decision that would allow mining to continue for several more years unregulated by zoning. It is difficult to argue that notice and a hearing should not be required when an agency makes a decision with the potential for such a significant and lasting impact.

But as with most judicial opinions that break new ground, *Calvert* leaves significant issues unresolved. Most significantly, it is unclear how extensive its impact will be. As will be discussed, *Calvert* could be interpreted to require numerous purportedly nonconforming mining operations in the state to undergo a notice-and-hearing process to confirm their vested rights status. It is not clear, moreover, whether the newly recognized notice-and-hearing requirement applies only to Diminishing Asset Doctrine determinations, or whether it also applies to other nonconforming use issues such as expansion and abandonment. If the latter is true, the number of nonconforming use matters subject to the new notice-and-hearing requirement could be numerous. Finally, it is unclear what applicability the California Environment Quality Act (CEQA) will have to vested rights determinations now that such determinations must be made following notice and a hearing. Although vested rights determinations have never previously been thought to be subject to CEQA, one could have argued the same about notice and a hearing before *Calvert*.

CEQA = California Environmental Quality Act

The *Calvert* Decision

In Calvert, *a community development director determined a mining operation was vested based solely on information the operator had provided. The county did not provide notice or hold a hearing before it made this decision.*

In *Calvert*, a large mining operator owned 3,430 acres of the aggregate-rich Yuba Goldfields, which it intended to mine under a claim of vested rights. To perfect this claim, the operator submitted to Yuba County, its lead agency, a six-page letter, 70-page memorandum of law and fact, and nearly 370 exhibits, all documenting the nature and scope of its historic activities. Without notice to the public or the mine's neighbors, and without holding a public hearing, the county's community development director informed the operator that, based on the information submitted, it had a vested right to mine all 3,430 acres. Later becoming aware of this determination, the petitioners—a neighbor and an activist organization—sued the county, claiming that the staff-level determination deprived them of procedural due process. The petitioners sought a writ of mandate to reverse the county's determination, compel it to provide appropriate notice, and re-determine the vested rights claim through a public hearing.

In deciding the issue, the court revisited the seminal Supreme Court case, *Horn v. County of Ventura*, 24 Cal. 3d 605 (1979). *Horn* held that an agency must provide notice and a hearing before making an adjudicative land use decision with the potential to substantially deprive the property interests of others. *Id.* at 612–616. Recognizing *Horn's* applicability, the *Calvert* court had to address two questions: (1) was the county's vested rights determination **adjudicative**, and, (2) if so, did the county's determination have the potential to substantially deprive the petitioners' property interests. The court answered yes to both questions.

Ministerial decisions are those that do not require the decisionmaker to exercise discretion. Adjudicative decisions, in contrast, involve the application of imprecise, subjective criteria.

Addressing the first issue, the court squarely rejected the notion that the determination was anything but adjudicative. The operator contended that the decision was **ministerial**, one in which the county had no discretion. Had the decision been ministerial, notice and hearing would not have been required. Similar to a building permit, the decision would have been made at the staff level, predicated only on whether the applicant had met certain objective criteria.

The court believed that the issues at hand involved much more. Focusing on SMARA's definition of **vested rights**, the court found that the criteria for determining whether a mining operation could lawfully operate as a nonconforming use were hardly objective. As Public Resources Code section 2776, provides:

> No person who has obtained a vested right to conduct surface mining operations prior to January 1, 1976, shall be required to secure a permit pursuant to [SMARA] as long as the vested right continues and as long as no substantial changes are made in the operation except in accordance with [SMARA]. A person shall be deemed to have vested rights if, prior to January 1, 1976, he or she has, in good faith and in reliance upon a permit or other authorization, if the permit or other authorization was required, diligently commenced surface mining operations and incurred substantial liabilities for work and materials necessary therefor.

In the court's view, the SMARA definition of vested rights and the requirements of the diminishing asset doctrine involve much more than the application of objective standards. The broad, imprecise standards they impose make nonconforming mining determinations adjudicative, rather than ministerial, decisions.

As the court observed, several terms in section 2776 involve "factual issues that must be resolved through the adjudicative exercise of judgment rather than the ministerial (automatic, nondiscretionary) application of fixed standards and objective measurements." *Calvert*, 145 Cal. App. 4th at 624. The operator must demonstrate that mining has been "contin[uous]" and that "substantial changes" have not been made. The operator must also demonstrate that work had been "diligently commenced" and "substantial liabilities" were incurred prior to adoption of the zoning ordinance that made the use nonconforming. Other vested rights issues require a similar exercise of judgment. Under the diminishing asset doctrine, for instance, the operator must demonstrate an objective manifestation of the intention to mine the area for which vested rights are claimed. In the court's view, all of these inquiries involved far more than the mere application of objective criteria; they required that a decisionmaker exercise discretion through application of broad, imprecise standards. *Id.* at 625.

In addition, the court analogized a vested rights claim to a conditional use permit. Although acknowledging the differences between the two, the court noted that, like permits, vested rights define the extent of a miner's rights to operate, making them the "fulcrum" governing the measures needed for a SMARA reclamation plan and financial assurance. *Id.* Because a conditional use permit must be issued following notice and a hearing, the court added, vested rights should be determined through a similar process. The differences between vested rights and a conditional use permit do not justify a different approval procedure for each. *Id.* at 625–626.

After rejecting the notion that a vested rights determination was ministerial, the court turned to the next question, whether the determination implicated a substantial deprivation of a property interest, triggering due process requirements. Based on the nature of the work, the court had little trouble answering this question in the affirmative. Factually, the operator had been mining on about 1,200 acres and claimed a right to mine another 3,400 acres. According to the court, "[t]his description itself is enough to envision significant environmental consequences and adverse effects to adjacent properties." *Id.* at 626. Not stopping there, the court went on to observe that one petitioner lived within 300 feet of the operation, subjecting him to noise, dust, and water pollution. The mining had also damaged aquatic species and created significant heavy truck traffic on local roads. All these impacts left no doubt that the potential impacts on the petitioner's interests were substantial, invoking the due process right to notice and a hearing. *Id.* at 627.

Because of the nature and extent of the mining in Calvert, *the court had little trouble concluding that the potential impacts on neighboring properties required notice and a hearing consistent with due process requirements.*

Finding a right to due process, the court held that the county's determination was invalid and ordered the operator to prove the vested rights claim at a public hearing following notice to affected landowners. The court did not specify the manner in which notice be provided, but suggested that an appropriate method would be to mail notices to property owners residing within a reasonable distance of the mine and post a notice at or near the site. *Id.* at 632. The court left other details to the agency to resolve.

SMGB = State Mining and Geology Board

As of this book's publication, the operator in *Calvert* was to submit the claim of vested rights at a hearing before the State Mining and Geology Board (SMGB), which had assumed the county's enforcement authorities under SMARA. *See* chapter 6, (describing SMGB takeover of lead agency authorities). At the time of this writing, the SMGB was in the process of adopting regulations governing notice-and-hearing procedures to be followed in this and other vested rights matters. A decision on the operator's vested status is likely in 2008.

Unresolved Issues Following *Calvert*

Calvert's holding seems logical. It makes sense that notice and a hearing should be required before an agency determines that a large mining operation could continue unregulated by zoning for many more years. But in establishing the basic requirements for notice and a hearing, *Calvert*—like any groundbreaking opinion—has opened up significant new issues. In an industry for which nonconforming uses are prevalent, the opinion could have a far-reaching effect.

Foremost among the unresolved issues is whether the decision is retroactive. Recognizing that this issue would arise, the court attempted to offer guidance:

Calvert may retroactively require numerous purportedly nonconforming mining operations to submit their vested rights claims to their lead agencies.

> We prefer to steer a middle course of limited retroactivity here, making our decision apply to all cases, including the one before us, in which no final judgment on appeal has yet been rendered, or in which an administrative determination of SMARA-based vested rights, in the context presented here of diminishing asset surface mining, is yet to be made or has been made and is still subject to administrative or judicial review.
>
> 145 Cal. App. 4th at 630.

While this passage confirms Calvert's retroactive effect as to matters then pending in court or before an administrative agency, it leaves untouched the question whether vested right issues never previously recognized must undergo the notice and hearing process the decision requires. Because many nonconforming mining operations are several decades old, agencies have often treated it as a given that the operations are lawfully nonconforming, without having made any formal determination as such. Although the precise number of these cannot be identified, the number could be significant—potentially numerous operations in several cities and counties throughout the state.

Taking the above passage at face value, however, it appears that *all* purportedly nonconforming operations, unless they have previously obtained a formal determination as to their vested status, are now required to undergo a notice-and-hearing process. This means that cities and counties, without any prompting, should require that mining operations submit their vested rights claims, unless they have previously done so, so that the claims may be confirmed following notice and a hearing. This is the case regardless of how long the agencies have considered the operations lawfully vested, how firmly held their beliefs may be as to the vested status, or whether the right to operate is a matter of public controversy. It remains to be seen whether agencies will do this; but a literal reading of *Calvert* would appear to presage a cottage industry for vested mining-right determinations.

A literal reading of Calvert would appear to presage a cottage industry for vested mining rights determinations.

Another issue left unresolved following *Calvert* pertains to the specific nonconforming use issues that must be decided following notice and a hearing. Reading from the court's holding, one could conclude that only diminishing asset doctrine issues need be decided. Summarizing its holding in the introductory portion of the opinion, the court appeared to suggest this, stating, "Our principal conclusion is that if an entity claims a vested right pursuant to SMARA to conduct a surface mining operation that is *subject to the diminishing asset doctrine,* that claim must be determined in a public adjudicatory hearing that meets procedural due process requirements of reasonable notice and an opportunity to be heard." *Id.* at 617 (emphasis added). Later in the opinion the court reiterated that the decision applies "only to an entity claiming a vested right under SMARA to conduct a surface mining operation *that is subject to the diminishing asset doctrine.*" *Id.* at 630 (emphasis added). Elsewhere, the court stated that its decision applies in cases "in which an administrative determination of SMARA-based vested rights, *in the context presented here of diminishing asset surface mining,* is yet to be made. . . ." *Id.* (emphasis added).

In light of the court's repeated focus on the diminishing asset doctrine, one might assume that only issues involving that doctrine require notice and a hearing. But the diminishing asset doctrine is only one of a number of issues that can arise in the context of a nonconforming mining operation. Moreover, the doctrine deals only with the geographic extent of the mine—how far an operator may expand extraction on the property where mining occurs. Diminishing asset issues are thus distinct from issues related to whether sufficient operations were underway at the time the restrictive ordinance was adopted, whether the claimed nonconforming operations had permissibly expanded—either because of an increase in market demand of because of the modernization of plants, equipment, and facilities—or whether the operations had at any time been abandoned. Strictly from the quoted passages, it would appear those issues are not subject to *Calvert's* notice-and-hearing requirements.

Language in Calvert *suggests that only diminishing asset doctrine determinations require notice and a public hearing. But reading the opinion as a whole, it is doubtful this is what the court intended.*

It is doubtful that this is what the court intended. At several places in the decision, the court recognized other issues that arise in the context of a nonconforming mining operation, referring several times to Public Resources Code section 2776. It noted that section 2776 focuses on "substantial changes" in the mining operation, the operator's "diligent commence[ment]" of operations, and the operator's incurrence of "substantial liabilities" prior to the adoption of the restrictive zoning ordinance. *Id.* at 624. These issues, the court added, "can only be determined on a case-by-case basis in a proper vested rights proceeding before the lead agency." *Id.* Referring repeatedly to "SMARA-based vested rights" or "vested rights under SMARA," the court further indicated that all of the section 2776 elements are subject to notice-and-hearing requirements. *Id.* at 630–631.

Later in the opinion, after noting that SMARA creates no procedure for recognizing vested rights, the court stated, "the *existence, nature and scope* of [vested] rights must be determined pursuant to some procedure even if SMARA fails to specify one. It goes without saying that that procedure must be a constitutional one." *Id.* at 629 (emphasis added). In a footnote, the court further declined to review "the parties' evidence and arguments regarding the *existence, nature and scope* of [the operator's] alleged vested right. . . ." *Id.* at 632 n.3 (emphasis added).

As it explained, that "will be the subject of the public adjudicatory hearing on vested rights. . . ." *Id.* Looking at the opinion as a whole, therefore, *Calvert*—while containing a number of references to just the diminishing asset doctrine—appears to require that *all* of the issues associated with a nonconforming mining operation be decided following notice and a hearing.

The significance of this aspect of the decision is perhaps greatest in regard to issues of expansion. Many of the nonconforming use issues arising in the context of mining deal with increases in production, addition of new plants and equipment, and modernization of facilities. Especially for activities that became nonconforming several years before, operators do not seek formal lead agency approval before implementing these changes. This is not because they wish to keep the changes secret; operators usually don't even recognize that the changes may constitute an impermissible expansion, or they believe in good faith that the changes are within their rights. But if *Calvert* applies—and this appears to be the case—then operators should now be required to seek confirmation of their right to effect these changes in a manner that comports with *Calvert's* notice-and-hearing requirement.

Because of the gray area separating permissible from impermissible expansions, however, an operator will have difficulty predicting when confirmation is necessary. The line separating the two is far from precise, raising the question: must an operator undergo a notice-and-hearing process for any change that might be deemed an impermissible expansion?

Calvert's *applicability to questions regarding expansions of nonconforming mining uses may present confounding issues for operators and lead agencies.*

And while *Calvert* appears to impose a duty to decide all expansion questions following notice and a public hearing, how will agencies comply with this aspect of the decision? Current information about mining operations is not always available, and agencies won't necessarily know when an impermissible change requires a hearing.

Perhaps the most perplexing question relates to CEQA's applicability now that vested rights issues must be decided through procedures similar to ones required for permits. Prior to *Calvert*, vested rights decisions did not have to be made in accordance with CEQA. But now that vested right decisions require a notice and a hearing, it is reasonable to ask whether other requirements applicable to permits—such as CEQA review—apply.

Perhaps the most perplexing issue following Calvert *is what applicability CEQA will have in vested rights determinations.*

As explained in chapter 8, CEQA clearly has a central role in a lead agency's review of mining permits, as well as related authorizations for a general plan amendment, zoning change, and reclamation plan. These authorizations constitute "discretionary projects" that trigger the need for environmental analysis under the Act. *See* Pub. Res. Code § 21080(a) (CEQA applies to "discretionary projects proposed to be carried out or approved by public agencies . . . ."); *Lexington Hills Ass'n v. State of California* (1988) 200 Cal. App. 3d 415, 430–433.

Following *Calvert*, it might be logical to conclude that the same is true for vested rights determinations. Indeed, their discretionary nature underpinned the court's reasoning. The potential for environmental impacts associated with those determinations was, in turn, central to the court's conclusion that the vested right determination at issue implicated due process rights of neighboring properties. Using this reasoning, one might plausibly assert that vested rights, like permits, involve the type of discretionary projects that trigger CEQA.

But strong reasons would militate in favor of a contrary conclusion. Vested rights and permits have similarities, but they also differ significantly. While each determination requires the exercise of discretion, an agency's authority to accept or reject vested rights is far more constrained than its power to approve or deny a permit. A permit can be denied for any number of reasons related to the promotion of public health, safety, and welfare. *See generally* chapter 7. By contrast, an agency considering a claim of vested rights is confined to deciding if the mining use lawfully existed at the time zoning became applicable, whether the mine operated continuously since that time, and whether the use has operated without expanding beyond its permissible scope. A operator who meets these standards has a *constitutional* right to continue mining. An agency cannot deny that right for reasons unrelated to these standards, such as a perception that the mining operation might create a nuisance or be incompatible with its surroundings.

Thus, while vested rights determinations involve discretion, the scope of an agency's discretion is far more limited than for a new mining application. It is questionable, then, whether a vested rights determination would constitute the type of "discretionary" act that triggers CEQA's applicability. *See* Pub. Res. Code § 21080(a). Certainly, an argument can be made in favor of that position. But it is worth noting that the discretion at issue in *Calvert* related to whether due process required notice and a hearing—an analytically distinct issue. Faced with the issue of CEQA's applicability, a future court may conclude that vested rights determinations do not involve the type of discretion sufficient to require environmental review under that act, even if they do implicate the due process of neighboring landowners.

Whichever interpretation prevails, this issue is likely to confound agencies and operators until it is litigated. Because decisions involving mining are increasingly becoming the source of public opposition, though, it may not take long for that to occur.

Part Four

Environmental Regulation

in this chapter...

chapter 10

Air Quality

Surface mining is an intensive industrial activity that affects the environment. Besides the required entitlements described in previous chapters, a mine must obtain numerous authorizations under federal, state, and regional environmental laws, many concerning air quality. A typical mine has a host of plants, equipment, and machines capable of emitting dust and other particulates. These include crushers, screeners, washing plants, and conveyors. Some operations have asphalt and cement plants onsite, while others have equipment for recycling concrete and asphalt. Unabated, the emissions from these services can be significant. For that reason, operators must obtain permits from regional or local air pollution control districts. Required by both federal and state law, the permits generally mandate special pollution control measures and abatement devices.

Clean Air Act

At the federal level, air quality is regulated by the Clean Air Act (CAA), which among other things establishes specific air quality standards, requires that states adopt plans to achieve those standards, and specifies the minimum criteria that certain new, major emitting air pollution sources must meet. *See generally* 42 U.S.C. § 7401 *et seq.* The CAA's regulatory scheme is federalist. While it imposes specific air quality mandates, the CAA leaves implementation to the states.

Air quality is regulated at the federal level under the Clean Air Act.

CAA = Clean Air Act

National Ambient Air Quality Standards

A principal objective of the CAA is to "protect and enhance the quality of the nation's air resources so as to promote the public health and welfare and the productive

The CAA establishes the NAAQS, which regulate air pollutants that can be reasonably anticipated to endanger public health and welfare.

EPA	=	Environmental Protection Agency
NAAQS	=	National Ambient Air Quality Standards
PM_{10}	=	Particulate matter
SIP	=	State Implementation Plan

capacity of its population." 42 U.S.C. § 7401(b)(1). Central to the fulfillment of this objective is the Act's establishment of the National Ambient Air Quality Standards (NAAQS). Through the NAAQS, the Environmental Protection Agency (EPA) has identified certain pollutants that "cause or contribute to air pollution which may reasonably be anticipated to endanger public health or welfare." *Id.* § 7408(a)(1)(A).

The NAAQS consist of several primary pollutants—ozone, carbon monoxide, lead, oxides of nitrogen, and particulate matter. The latter, also known as **PM_{10}** is the pollutant that generally subjects surface mining to the CAA as well as the state and regional air quality regulations discussed below. Technically speaking, PM_{10} is particulate matter with an aerodynamic diameter of 10 or fewer microns. Defined less technically, PM_{10} is made up of fine, breathable particles. Prolonged human exposure to it can create significant health problems—including aggravated asthma, chronic bronchitis, decreased lung function, and even premature death. Bay Area Air Qual. Mgt. Dist., *Particulate Matter* (**www.baaqmd.gov/dst/pollutants.htm#pm**).

Under the CAA, areas that comply with the NAAQS standard for a pollutant are considered attainment areas; those areas that do not comply are considered non-attainment areas.

For PM_{10} and other prescribed pollutants, the NAAQS determine whether designated regions of a state may be considered **attainment** or **non-attainment** areas. The EPA has developed thresholds for each regulated pollutant. A region that exceeds the threshold for a particular pollutant is considered a non-attainment area. 42 U.S.C. § 7407(d)(1)(A)(i). By contrast, one that does not exceed the threshold is considered an attainment area. *Id.* § 7407(d)(1)(A)(ii).

State Implementation Plans

Every state must adopt a SIP including emission limitations and a regulatory program for construction of stationary sources.

To achieve compliance with all NAAQS, every state must adopt and receive EPA approval for a state implementation plan (SIP). 42 U.S.C. § 7410. A SIP must provide for the "implementation, maintenance, and enforcement" of the NAAQS. *Id.* § 7410(a)(1). It must "include enforceable emission limitations and other control measures, means, or techniques... to meet the applicable requirements of [the CAA]." *Id.* § 7410(a)(2)(A). In addition, a SIP must include a program for "regulation of the modification and construction of any stationary source within the areas covered by the plan as necessary to assure the [NAAQS] are achieved...." *Id.* § 7410(a)(2)(C).

New Source Review for Major Emitting Facilities in a Non-Attainment Area

The CAA mandates that states require permits for major stationary sources in non-attainment areas.

A permitting program is yet another critical component of a SIP. 42 U.S.C. § 7410(a)(2)(c). Under the CAA state and local governments have primary responsibility for control of air pollution. *Id.* § 7401(a)(3). To meet the Act's overall goals, a state must require a permit for the construction and operation of new or modified "major stationary sources" in a non-attainment area. *Id.* § 7502(c)(5); *see also id.* §7411(a)(3) (defining "stationary source" as any "building, structure, facility, or installation which emits or may emit any air pollutant").

Whether a stationary pollutant source is major depends on the degree of an area's non-attainment with a particular pollutant. For PM_{10}, non-attainment areas are classified either as **moderate** or **serious**. *Id.* § 7513. In a moderate non-attainment area, a major source is one that directly emits, or has the potential to emit, 100 tons or more of PM_{10}. *Id.* § 7602(j). For a serious area, a major source emits, or has the potential to emit, 70 tons or more. *Id.* § 7513a(b)(3).

BACT = Best available control technology
LAER = Lowest available emission rate
NSR = New source review
PSD = Prevention of significant deterioration

Under the CAA, a major stationary source is subject to "new source review" (NSR). Under NSR requirements, any person proposing to construct or operate a major new stationary source, or proposing to modify that source, must comply with two principal requirements. First, the source must achieve the "lowest available emission rate" (LAER). *Id.* § 7503(a)(2); *Id.* § 7411(a)(4) (defining modification of a source as "any physical change in, or change in the method of operation of, a stationary source which increases the amount of any air pollutant emitted...."). This means that the source must achieve the more demanding of either: (a) "the most stringent emission limitation which is contained in [the SIP] for such class or category of source....;" or (b) "the most stringent emission limitation which is achieved in practice by such class or category of source...." *Id.* § 7501(3).

When subject to NSR, a stationary source must achieve LAER.

Second, the source must satisfy offset requirements. Specifically, by the time the operation has begun, the source may not have increased overall emissions in the non-attainment region beyond what existing sources—and proposed or modified non-major sources—generated at the time of the permit application. *Id.* § 7503(a)(1)(A).

Stationary sources subject to NSR must satisfy offset requirements.

Prevention of Significant Deterioration for Major Emitting Facilities in Attainment Areas

A major emitting facility in an attainment area is also subject to permitting requirements under the CAA. For surface mining, a **major emitting facility** is any Portland cement or phosphate rock plant that produces 100 tons or more per year of any NAAQS pollutant; other mining uses that produce 250 tons per year of NAAQS pollutants are considered major emitting facilities. 42 U.S.C. § 7479(1). These facilities are subject to CAA's **prevention of significant deterioration** (PSD) provisions. *Id.* § 7471.

Major emitting facilities in attainment areas are subject to the PSD requirements of the CAA.

A facility subject to PSD is generally entitled to emit pollutants at a higher level than one in a non-attainment area. To obtain a permit, however, the facility may not exceed prescribed emission increments and cause or contribute to pollution that exceeds the NAAQS standard for a particular pollutant. *Id.* § 7475(a)(3). The facilities are subject to **best available control technology** (BACT) that, "taking into account energy, environmental, and economical impacts and other costs," the permitting authority determines "is achievable for such facility through application of production processes and available methods, systems, and techniques...." *Id.* §§ 7475(a)(4), 7479(3). BACT for PSD facilities is determined on a case-by-case basis. *Id.* § 7479(3).

When subject to PSD, major emitting facilities must achieve BACT.

California's Air Regulatory Program

While the CAA requires that every state adopt a SIP and mandates that major facilities obtain permits meeting certain criteria, California's air regulatory program goes well beyond the minimum CAA standards.

General Requirements

To comply with CAA standards, California has adopted numerous statutes, regulations, and rules that collectively constitute its SIP. While state government shares responsibility for controlling air pollution under the SIP, the primary authority for controlling pollution from non-vehicular, or stationary, sources rests with regional and local authorities. Health & Safety Code § 39002; *see also id.* § 39001 ("[s]ince air pollution knows no political boundaries, the Legislature declares that a regional approach to the problem should be encouraged whenever possible. . . .") As part of its air regulatory program, California has divided the state into regional air basins. *Id.* § 39001. Each basin is designated as either an attainment or non-attainment area for each regulated pollutant. *Id.* § 39608(a).

California has divided the state into regional air basins for purposes of determining pollutant attainment status.

To be considered an attainment area for a particular pollutant, each basin must not exceed the federal NAAQS threshold. However, California has adopted its own ambient air quality standards, and a basin may not exceed state standards that are, in many cases, more rigorous. *Id.* § 40001; *see generally* Cal. Code Regs. tit. 17, § 70101 *et seq.* (setting forth California's ambient air quality standards). A basin may comply with the federal PM_{10} standard, for instance, if its measured concentration is equal to or less than 150 micrograms per cubic millimeter. 40 C.F.R. § 50.6(a). By contrast, to comply with California's stricter standard, a basin may not exceed a measured concentration of 50 micrograms per cubic millimeter. Cal. Code Regs. tit. 17, § 70200.

California has adopted its own standards that in many instances are more stringent than the NAAQS.

CARB = California Air Resources Board

The California Air Resources Board (CARB) is responsible for division of the state into several air quality basins. Health & Safety Code § 39606(a). CARB sets the ambient air quality standards and is responsible for statutes and regulations that govern air pollution control. *Id.* §§ 39601, 39606. As noted above, non-vehicular pollution control is primarily a regional responsibility. *Id.* §§ 39001, 39002. California thus regulates air pollution through a variety of regional bodies, including air quality management districts, county air pollution control districts, and unified air pollution control districts.

Although CARB has adopted California's air quality standards and has divided the state into basins, regional bodies regulate air pollution within the state.

Each district, in turn, is responsible for its own implementation plan that, when approved by CARB, becomes part of the California SIP. *Id.* §§ 40000, 41650–41652. A district is also responsible for ensuring that all federal and state ambient air quality standards are met. *Id.* § 40001(a). In fulfilling its responsibilities, a district is not limited to ensuring compliance with federal and state standards. It may also impose more stringent standards through its own regulations, as many districts have done. *Id.* §§ 39002, 41508.

Air pollution control districts must adopt their own implementation plans to meet federal and state air quality standards.

Air Pollution Control District Permitting Schemes

A central component of district air pollution regulation is a permitting scheme that, as noted above, is mandatory under the CAA for certain major emitting facilities. A permitting scheme may exceed federal requirements, but it is through a permitting scheme that a district complies with its responsibilities under the NSR and PSD provisions of the CAA.

A central component of district air pollution regulation is a permitting scheme.

In mandating permits, a district usually requires two types of authorizations: "authorities to construct," regulating the construction of machines, equipment, and facilities; and "permits to operate," governing pollution emissions during the operation of the machines, equipment, and facilities. *See, e.g.*, Bay Area Air Qual. Mgt. Dist. Regs. §§ 2-1-301, 2-1-302. At a minimum, a district may approve the permits only if issuance would not "prevent or interfere with the attainment or maintenance of any applicable air quality standard" and will comply with all federal, state, and district rules and regulations. Health & Safety Code § 42301(a), (b).

Air pollution control districts usually require two types of permits: authorities to construct and permits to operate.

District requirements for new source review. District permit requirements usually do more than the CAA requires. While the Act requires that a major new stationary source in a non-attainment area achieve the LAER standard of pollution control and satisfy offset requirements, the NSR provisions are not triggered unless certain emission threshold levels are met. In California, districts often reduce that threshold below what federal law mandates. In the South Coast Air Quality Management District, for instance, all new sources are subject to NSR if they "may cause the issuance of any non-attainment air contaminant. . . ." S. Coast Air Qual. Mgt. Dist. Rule 1301(b)(1).

Some California air pollution control districts subject more sources to NSR than federal law requires.

For a facility subject to new source review, a district may require that proposed or modified sources achieve BACT, a standard similar to the CAA definition of LAER. A district will typically define BACT as follows:

District NSR usually subjects sources to BACT, which is the equivalent of the federal LAER standard.

- The most effective emission control device or technique that has been successfully utilized for the type of equipment comprising such a source
- The most stringent emission limitation achieved by an emission control device or technique for the type of equipment constituting the source
- Any emission control device or technique determined to be technologically feasible and cost-effective
- The most effective emission control limitation for the type of equipment comprising such a source, which the EPA states . . . is contained in an approved implementation plan of any state, unless the applicant demonstrates to the satisfaction of [the air pollution control officer] that such limitations are not achievable

Bay Area Air Qual. Mgt. Dist. Rule 2-2-206

As this definition makes clear, BACT generally focuses on operating techniques and control devices placed on polluting equipment, machines, and facilities. Typical techniques and devices for mining operations are discussed below.

District NSR requirements also mandate that new or modified sources comply with offset requirements that preclude any net emission increase. In some cases, an offset will be required only for sources that meet the "major emitting facility" standard of the CAA—generally 250 pounds of a NAAQS pollutant per year. When subject to offset, the required amount may vary depending upon a particular district's rules and the location of the emitting party's offsets. The Sacramento Metropolitan Air Quality Management District, for example, requires that an operator provide a 1:1 ratio of offsets from the same source, but that those from other sources be provided in 1.2:1 or 1.5:1 ratios. Sacramento Met. Air Qual. Mgt. Dist. Rule 202-303.

District NSR schemes often impose other requirements such as metering, monitoring, and reporting.

To ensure compliance with its permitting responsibilities, a district may mandate installation of metering and monitoring devices, and will frequently impose strict reporting requirements. A district may likewise impose severe penalties for tampering with monitoring devices or inaccurate reporting.

District PSD schemes may exceed the requirements of federal law.

District requirements for prevention of significant deterioration. District permitting schemes also implement PSD provisions of the Clean Air Act. As with NSR, the district may establish standards that exceed federal law. Generally, new sources subject to PSD must conform with the district's best available control technology. *See, e.g.*, S. Coast Air Qual. Mgt. Dist. Rule 1703(a)(2). However, they are not subject to offset. Instead, districts usually have models showing that emissions from new sources, along with those from all stationary sources in a baseline area, will not cause or contribute to air pollution that exceeds any ambient air quality standard or allowable pollution increment over an area's baseline concentration. *See id.* § 1703(a)(3)(C).

Approval and Enforcement of Air Quality Regulations

Obtaining an air pollution control permit. Under the standards described above, a mine operator intending to install plants, equipment, or machinery that will discharge air contaminants must first obtain district authorization. Typically, an operator must obtain approval of "authorities to construct" and "permits to operate." *See, e.g.*, Bay Area Air Qual. Mgt. Dist. Rules 2-1-301, 2-1-302.

Districts must comply with a number of measures to expedite their review of air pollution permit applications.

To expedite district review and approval of air quality permits, a district must adopt a list of permitting criteria that identifies streamlined requirements for a particular project, such as a new surface mining operation. Health & Safety Code § 42301.3(c). The list may include "specific information regarding the pollution control equipment or operational changes that will reduce emissions." *Id.* The legislature has also enacted the Permit Streamlining Act to ensure expedited review for air quality permits. The Act requires that each district institute programs to "reduce unnecessary delay." *Id.* § 42322(a). Among the features for expedited review, a district must establish pre-certification for mass-produced equipment operated by many owners under similar conditions. *Id.* § 42322(a)(1). Expedited review must also include a consolidated permitting program for any source requiring more than one permit, so that all aspects of a particular operation can be reviewed at once. *Id.* § 42322(a)(2).

The legislature has also established other measures designed to expedite permit review and approval. *Id.* § 42322(a)(3)–(7). California law specifically states that, as long as an applicant provides all required information, a district complete its review and deem it complete within 30 days. *Id.* § 42301.3(d)(1).

Enforcement. State law imposes several consequences for violating air pollution control statutes and regulations, including criminal penalties. Health & Safety Code § 42400(a); *see, e.g., id.* § 42400.2(a) (providing that person knowing of emission of an air contaminant violating state or district regulations who fails to take corrective action may be fined up to $40,000 and sentenced up to one year in jail); *Id.* § 42400.3(a) (stating that anyone who intentionally emits an air contaminant violating state law or district regulations may be fined up to $75,000 and sentenced up to one year in jail). Civil penalties may also be imposed. *See id.* § 42402(a), (b); *see, e.g., id.* § 42402.3 (mandating a civil penalty of up to $75,000 on one who intentionally emits an air contaminant in violation of state law or district regulations). Civil actions or district or CARB administrative proceedings may seek further penalties. *Id.* §§ 42402.5, 42404, 42410.

State law imposes a number of criminal and civil penalties for violation of air quality standards.

A district may also issue an abatement order to anyone who violates state or district laws and regulations. *Id.* § 42451(a). Upon failure to comply, a district may seek an injunction against the person in superior court. *Id.* § 42453. That the order has been issued and that the violator has not complied "shall be sufficient proof to warrant the granting of a preliminary injunction." *Id.* § 42454.

Air pollution control districts can seek injunctive relief against persons or entities that violate air quality standards.

Control of Air Pollution from Surface Mining

Identifying every specific pollution control measure a mine must employ to comply with the regulations discussed above is not possible. The requirements for specific sites will vary depending on the type of mining, its location, and attainment status, and any particular standards the governing air pollution control district has adopted. Nonetheless, air quality permits for mining operations can impose common restrictions, including some or all of those described below.

Common Air Pollution Abatement Devices and Practices

BACT, a central component of air pollution control district permitting requirements discussed above, regulates the devices an operator must use for plants, equipment, and facilities and the pollution control practices that must be followed. In implementing BACT for mining, a district usually specifies that the operation use certain pollution abatement devices.

To meet BACT, mining operators often must install particulate-control devices on their plants and equipment.

A district permit, for example, may require the use of a **baghouse**. Operating like the bag on a vacuum cleaner, a baghouse traps large particles freed by mining plants and machinery. Or the permit may state that the plant must be equipped with a **cyclone** that uses centrifugal force to remove particles from polluted air. Some permits require a **wet scrubber**, a device that utilizes high energy water or reactant spray to remove pollutants, either by absorption or reaction. An operator using crushing equipment may need a water spray system to eliminate fugitive dust migration. *See generally*, *e.g.*, Bay Area Air Qual. Mgt. Dist., *Permit Handbook*, § 11, Crushing and Grinding.

Districts often require that mining operators water stockpiles and onsite areas and control track out onto public roads.

Some districts may regulate dust migration from stockpiles and on-site roads. An operator may have to water stockpiles and unpaved road surfaces periodically, sometimes with chemical additives as well as water. The rules might also mandate that an operator must control **track out** of dust and debris onto public roads by making a loaded truck pass over or through rumble grates, wheel washers, or truck washers. *See*, *e.g.*, S. Coast Air Qual. Mgt. Dist. Rule 1157(d)(7)–(8).

Nuisance and Visible Emissions Regulations

Health and Safety Code section 41700 prohibits mining operations from emitting nuisance emissions.

Air pollution control district permits further regulate mining operations by prohibiting nuisance emissions. Health and Safety Code section 41700, setting forth the **nuisance limitation**, prohibits any person from discharging:

> [F]rom any source whatsoever such quantities of air contaminants or other material which cause injury, detriment, nuisance, or annoyance to any considerable number of persons or to the public, or which endanger the comfort, repose, health, or safety of any such persons or to the public, or which cause, or have a natural tendency to cause, injury or damage to business or property.

Health and Safety Code section 41701 prohibits mining operators from emitting pollutants that exceed certain Ringelmann Test thresholds.

A more specific emission control is the visibility limitation required by Health and Safety Code section 41701, which precludes any person from emitting:

> [F]rom any source whatsoever any air contaminant... for a period or periods aggregating more than three minutes in any one hour which is: (a) [a]s dark or darker in shade as that designated as No. 2 on the Ringelmann Chart..., or (b) [o]f such opacity as to obscure an observer's view to a degree equal to or greater than does smoke described in subdivision (a).

The "Ringelmann" limit this provision adopts refers to a special test for measuring smoke and particulate emissions. It restricts visible emissions from being light gray for more than three minutes in any hour.* Some districts follow an even

* The Ringelmann test is a long-used test the United States Bureau of Mines adopted for controlling smoke and particulate emissions. To conduct the test, the observer holds up square cards in a line of site with distant emissions. The observer then compares the color of the emissions to the color shown by the cards, which contain five grids of progressively darker gradations. When viewed as the test requires, the cards will reveal shades ranging from white

stricter standard. In the Bay Area Air Quality Management District, no person may emit any emission exceeding Ringelmann No. 1 for more than three minutes per hour, which restricts visible emissions to being no darker than very light gray for that period. Bay Area Air Qual. Mgt. Dist. Rule 6-301.

Optional Registration Program for Portable Mining Equipment

Many mines use portable crushers, screens, and other movable equipment so that processing areas can be moved as excavation progresses to new surface areas. In addition, permanent plants can be prohibitively expensive.

An operator may treat portable equipment like a stationary source, obtaining permits from the governing air pollution control district by following the procedures described above. However, portable equipment can be moved from one district's territory to another, with differing district standards making air quality compliance difficult. To eliminate the burden of overlapping, inconsistent regulation—and the need to obtain separate permits from each district—the legislature has created an optional program allowing an operator to register portable equipment. *See* Health & Safety Code § 41750(a)–(c); Cal. Code Regs. tit. 13, § 2451(a).

To avoid overlapping and burdensome requirements, mining operators can register their portable equipment through a special registration program.

To implement the registration program, CARB has adopted regulations setting forth the process. *See generally* Health & Safety Code § 41752; Cal. Code Regs. tit. 13, §§ 2450–2465. For mining uses, eligible equipment includes portable machines used in sand and gravel screening, rock crushing, pavement crushing, and recycling operations. Cal. Code Regs. tit. 13, § 2457(b)(3). To register, an operator must submit an application to the Executive Officer of CARB, who may test each piece of equipment to ensure that prescribed air pollution control standards are met. *Id.* §§ 2454(b), 2460(a).

To register their equipment, mining operators must submit applications to the Executive Officer of CARB, who tests the equipment to ensure its compliance with air quality standards.

Under the standards for portable mining equipment, a machine may not emit more than ten tons per year of PM_{10} and 82 pounds per project of PM_{10}. *Id.* § 2457(a)(1)–(2). Nor may it cause the emission of hazardous pollutants—for example, those caused by a rock crushing plant operating at a serpentine quarry. *Id.* § 2457(d). The equipment must further meet the following standards:

- It cannot discharge air contaminants for more than three minutes per hour exceeding Ringelmann No. 1 or equivalent 20 percent opacity
- No emissions shall be visible beyond the operator's property line
- Transfer points and crushing equipment must (1) be ducted through a fabric or cartridge collector, or (2) be equipped with a wet suppression system

to black, with card zero being pure white, cards one through four being darkening shades of gray, and card five being pure black. Based on the number of measurements per the testing period, the observer concludes the test by assigning a number to the emission based on which card most closely resembles the color of the emissions. *See generally* U.S. Bureau of Mines, *Ringelmann Smoke Chart, Informational Circular* 8333 (1967).

- All conveyors must be covered, unless the transported material would not create visible emissions
- All stockpiled material from the portable unit must be maintained at a minimum moisture content unless the stockpiled material results in no visible emissions

Id. § 2457(b)(3)

Once an operator has registered equipment, it may operate the equipment anywhere in California without obtaining permits from the district to which the equipment is moved.

Once registered, the portable equipment may be used anywhere within California without needing a district permit. *Id.* § 2454(c). However, an operator moving that equipment into a district's territory for more than five days must notify the governing air pollution control district in writing. *Id.* § 2459(a). The operator must also comply with record-keeping and reporting requirements. *Id.* § 2458. In addition, CARB may require testing of the registered equipment up to once every three years. *Id.* § 2460(b).

Mining in Areas Containing Naturally-Occurring Asbestos

Long-term exposure to asbestos fibers released into the air can lead to asbestosis, mesothelioma, and lung cancer.

Special pollution control regulations apply where mining occurs in areas that contain naturally occurring asbestos. Enacted following 2002 legislation, the regulations address the serious health ailments created by long-term exposure to asbestos fibers. Scientific studies have determined that prolonged exposure can lead to lung cancer, mesothelioma, and asbestosis.

Operations that mine in certain ultramific rock deposits are subject to special asbestos regulations.

Because asbestos occurs naturally in ultramafic rock deposits, which include serpentine, the regulations only apply to certain surface mining operations. Cal. Air Res. Bd., *Regulatory Advisory, Asbestos Airborne Toxic Control Measure for Construction, Grading, Quarrying, and Surface Mining Operations* 1 (Jul. 29, 2002). Surface mining generally does not occur in ultramafic rock, so most operations are not subject to asbestos regulations. Cal. Code Regs. tit. 17, § 93105(b).

Operators subject to the asbestos regulations must obtain approval of an asbestos dust mitigation plan that demonstrates compliance with a number of standards.

When the regulations do apply, an operator must obtain approval of an asbestos dust mitigation plan from the governing air pollution control district. *Id.* § 93105(f)(1). That plan must demonstrate that the mining operations will meet all of the following requirements:

- The operation will not emit dust that is visible crossing the property line
- Crushing equipment will not discharge visible emissions exceeding 75 percent of Ringelmann 1 or 15 percent opacity for more than 3 minutes in any one hour
- Grinding mills, screening operations, and transfer points on conveyors will not discharge visible emissions exceeding 50 percent of Ringelmann 1 or 10 percent opacity for more than 3 minutes in any one hour
- To show that the above three conditions will be met, operators must undertake measures such as installing spray bars on all conveyors and installation of shrouds at drop points
- Inactive stockpiles prone to mechanical and wind disturbances will be wetted, covered, or controlled with palliatives, suppressants, paving, or wind berms or brakes

- On-site unpaved roads, parking lots, and staging areas will be stabilized by wetting, dust palliatives or suppressants, paving, or covering with gravel
- Quarried materials are adequately wetted
- All truck loads are adequately wetted before or during wetting and at the time they leave the site
- Vehicle speeds within the operation will not exceed 15 miles per hour
- Track out control measures, such as gravel pads, paving, or tire-washing systems will be installed, and track out onto public roads will be removed by a street sweeper

Id. § 93105(f)

In addition, the operator must comply with record-keeping and reporting requirements. *Id.* § 93105(f)(5). The air pollution control district may also mandate that an operator engage in periodic air monitoring. *Id.* § 93105(f)(3)(B).

in this chapter...

chapter 11

Water Quality

As with air, surface mining can pollute water. When removed from a mine pit or used for washing and processing, water carries chemicals, contaminants, or constituents that diminish the quality of groundwater and harm adjacent streams, rivers, and lakes. Runoff from overburden or waste piles can have a similar effect. Mining for metallic ore exposes certain compounds to air and water, resulting in acid drainage and seepage of harmful contaminants. Given its potential for water pollution, surface mining is subject to extensive regulation that involves compliance with numerous federal, state, and regional requirements.

Clean Water Act

CWA = Clean Water Act

As with air quality regulation, water quality law is federalist in nature, requiring states to implement and enforce federal broad policies. Under the Clean Water Act (CWA), the overall goal is that "discharge of pollutants into the navigable waters [of the United States] be eliminated...." 33 U.S.C. § 1251(a)(1). Because this goal has not yet been achieved, federal law in the interim seeks a goal that, "wherever attainable,... provides for the protection and propagation of fish, shellfish, and wildlife and provides for recreation" in waters of the United States. *Id.* § 1251(a)(2).

The CWA seeks a goal that, where attainable, provides for protection and propagation of fish, shellfish, and wildlife and provides for recreation.

Mandatory State Water Quality Policies Required

Under the CWA, every state must adopt and implement water quality standards similar to what the Clean Air Act requires for air. 33 U.S.C. § 1313(a)(1). At a minimum, a state's water quality standards must specify the uses that need to be achieved and protected. 40 C.F.R. § 131.10(a). This classification of uses must further:

To achieve its goals, the CWA requires that states adopt and implement water quality standards.

> [T]ake into consideration the use and value of water for public water supplies, protection and propagation of fish, shellfish, and wildlife, recreation in and on the water, agricultural, industrial, and other purposes including navigation. In no case shall a State adopt waste transport or waste assimilation as a designated use for any waters of the United States.
>
> *Id.*

The CWA requires that states adopt anti-degradation policies to maintain and protect water quality.

In addition, the CAA requires the adoption of water quality criteria to protect uses designated by the state. *Id.* § 131.11(a). The criteria must "be based on sound scientific rationale and... contain sufficient parameters or constituents to protect the designated use." *Id.* States must also adopt an **anti-degradation** policy ensuring the maintenance of a water quality level necessary to protect existing uses. *Id.* § 131.12(a)(1). Where the quality exceeds federal standards, however, a state's anti-degradation policy may allow lower quality if necessary to "accommodate important economic or social development in the area in which the waters are located." *Id.* § 131.12(a)(2). In doing so, the state must ensure that the water quality will be adequately protected. *Id.*

EPA = Environmental Protection Agency
NPDES = National Pollutant Discharge Elimination System

National Pollutant Discharge Elimination System

The NPDES requirements of the CWA prohibit the discharge of pollutants into navigable waters from any point source without a permit.

The NPDES program regulates pollutant discharge principally through effluent limitations.

The CWA implements its broad goals through various permitting schemes, principally the National Pollutant Discharge Elimination System (NPDES). The NPDES prohibits any person from discharging any **pollutant** from a **point source** into **navigable waters** without a permit. 33 U.S.C. §§ 1311(a), 1311(b)(1)(A), 1312(a), 1342(a)(1). Under the CWA, "pollutant" is defined as "the man-made or man-induced alteration of the chemical, physical, biological and radiological integrity of water." *Id.* § 1362(19). A "point source" is "any discernable, confined and discreet conveyance... from which pollutants are or may be discharged." *Id.* § 1362(14). "Navigable waters" are broadly defined as "waters of the United States." *Id.* § 1362(7).

California administers the NPDES program through issuance of waste discharge requirements.

Under the CWA, states that meet certain standards and obtain approval may administer the NPDES permitting program. *Id.* § 1342(b). Because a comprehensive state water quality program was in place before the CWA was adopted, California easily obtained Environmental Protection Agency (EPA) approval to administer the program, which it does by issuing waste discharge requirements. *See* Water Code § 13377. To ensure continuing compliance, these authorizations—sometimes referred to as NPDES permits/waste discharge requirements—are processed in accordance with CWA requirements and must meet CWA's substantive standards. Cal. Code Regs. tit. 23, §§ 2235.1, 2235.2.

California's regulation of water quality is broader than the federal scheme, regulating any discharge of waste and all waters within the state.

California's regulation of water quality is broader than federal law. The CWA applies only to discharges from "point sources" into "waters of the United States." By contrast, under California's scheme any discharge of **waste** is regulated, regardless of origin. Water Code § 13050(d). It is also broader in scope. The "waters of the United

States" subject to CWA regulation exclude many water bodies because the power of Congress to regulate commerce is limited by the constitution. *See generally Solid Waste Agency of N. Cook County v. U.S. Army Corps. of Eng'rs* (2001) 531 U.S. 159. Unconstrained by that, California's scheme encompasses all waters within the state's boundaries, whether public or private. (1966) 48 Ops. Cal. Att'y Gen. 30, 34–35. Significantly, this includes surface water as well as ground water, which the CWA does not regulate. Water Code § 13050(e).

California's Porter-Cologne Act and Water Quality Regulatory Scheme

California Water Plan

Even before the CWA first mandated minimum state water quality standards, California had already adopted a comprehensive water quality control scheme through the Porter-Cologne Water Quality Control Act of 1969. The aims of Porter-Cologne are far reaching, mandating that:

California regulated water quality before the CWA through the Porter-Cologne Water Quality Control Act of 1969.

> Factors which may affect the quality of the waters of the state shall be regulated to attain the highest water quality which is reasonable, considering all demands being made and to be made on those waters and the total values involved, beneficial and detrimental, economic and social, tangible and intangible.
>
> Cal. Water Code § 13000

To implement these broad goals, the state has adopted the California Water Plan, comprised of a variety of documents guiding the protection, conservation, development, and utilization of the state's water resources. *Id.* § 10005(a). Although required by federal law, California's plan goes well beyond the requirements of the Clean Water Act.

The state has adopted the California Water Plan, which provides principles, guidelines, and objectives for protecting the quality of its waters.

Under Porter-Cologne, the California Water Plan must include policies adopted by the State Water Resources Control Board. *Id.* §§ 13140–13141. Specifically, the policies must address "quality principles and guidelines for long-range resource planning, including ground water and surface water management programs...." *Id.* § 13142(a). State water quality must also address "[w]ater quality objectives at key locations for planning and operation of water resource development projects and for water quality control activities." *Id.* § 13142(b). The principles, guidelines, and objectives of state policy must further "be consistent with the state goal of providing a decent home and suitable living environment for every Californian." *Id.* § 13142.

As required by the CWA, California also has an anti-degradation policy. According to state law, when the quality of water exceeds the qualities required by the state, the existing high quality will be maintained. Cal. Water Res. Control Bd., *Statement of Policy With Respect to Maintaining High Quality Waters in California*, Resolution 68-16. Exceptions are permitted only when a change in water quality is consistent with maximum benefit to the people of the state, does not unreasonably affect present

California has adopted its own anti-degradation policy, which requires the maintenance of the existing high quality of water exceeding state standards.

and anticipated beneficial uses of water, and does not result in water quality that falls below state policy goals.

Regional Water Quality Control Plans

RWBs must adopt regional water quality control plans that identify beneficial uses and specify water quality objectives.

Just as California's Air Regulatory Program divides the state into regions and utilizes regional bodies for implemention and enforcement, California has divided the state into separate water basins with regional water quality control boards (RWBs). Each RWB is responsible for adopting its own regional water quality control plan, frequently called a **basin plan**. Water Code. § 13240.

BMP = Best management practice
RWB = Regional water quality control board
SWB = State Water Resources Control Board
WDR = Waste discharge requirement

At a minimum, a basin plan must identify **beneficial uses** for all waters within its region. *Id.* § 13241. Beneficial uses include "domestic, municipal, agricultural and industrial supply; power generation; recreation; aesthetic enjoyment; navigation; and preservation and enhancement of fish, wildlife, and other aquatic resources or preserves." *Id.* §13050(f). A basin plan must further identify **water quality objectives** that "ensure the reasonable protection of beneficial uses and the prevention of nuisance." *Id.* § 13241.

Waste Discharge Requirements

California regulates the discharge of wastes that could affect water quality through waste discharge requirements (WDRs). These implement the federal NPDES program as well as regulate the broader classes of waste covered by the Porter-Cologne Act.

Before discharging waste, an entity must first file a report of waste discharge with its RWB.

Reporting requirements and approval of waste discharge requirements. In accordance with Porter-Cologne, an entity proposing to discharge waste must first file a report with the governing RWB. Proposed discharges subject to the federal NPDES permitting program require a similar report. Water Code § 13376. Besides indicating the type of facility operated and the nature of the discharge, the report must be accompanied by an additional report detailing any physical and chemical characteristics that might cause contamination and information about prescribed testing. *Id.* § 13260(k)(1). The proposed discharger must also pay certain fees. *Id.* § 13260(d).

In prescribing discharge requirements, the RWB considers the beneficial uses to be protected, water quality objectives, and need to prevent nuisance.

Before approving or denying WDRs, the RWB must hold a hearing. *Id.* § 13263(a). To ensure consistency with the state and regional water quality control plans, the RWB must consider "the beneficial uses to be protected, the water quality objectives reasonably required for that purpose, other waste discharges, the need to prevent nuisance" and other prescribed factors. *Id.* The RWB may prescribe requirements in accordance with all these considerations. *Id.*

Unlike a traditional permitting scheme, the RWB cannot specify the manner of compliance. *Id.* § 13360(a). A discharger may comply in any lawful manner. *Id.* Yet, as a practical matter, a discharger should act voluntarily according to **best management practices** (BMPs) that the RWB or State Water Resources Control Board (SWB) has identified. These BMPs and related effluent limitations frequently become more strict as regulations in water quality control plans become more demanding. Once issued, a waste discharge requirement does not include a vested right, as a WDR is an expressly made "privilege," not a right. *Id.* § 13263(g). Generally, moreover, a WDR has a term of five years. *Id.* § 13378.

WDRs are privileges, not rights.

Waiver of waste discharge requirements. In some cases, an RWB may waive waste discharge requirements. The RWB may do so when it determines that a proposed discharge does "not pose a significant threat to water quality" or when the discharge would be consistent with applicable water control plans and the waiver would be in the public interest. Water Code § 13269(a)(1), (3). The RWB may make the waiver conditional, however, requiring among other things that the discharger comply with monitoring requirements. *Id.* § 13269(a)(2). A waiver may also be general, applying to a broad class or an individual discharge. The RWB or SWB may further terminate the waiver at any time. *Id.*

The RWB may waive WDRs when a proposed discharge does not pose a significant threat to water quality.

Enforcement. Porter-Cologne creates a variety of enforcement options when a person discharges waste without having first obtained a WDR. The RWB may issue a cleanup and abatement order, seeking injunctive relief when a discharger fails to comply. *Id.* § 13304(a). When the RWB requests that a report be filed and a discharger fails to do so, a violator may be guilty of a misdemeanor and liable for civil penalties. *Id.* § 13261. In that situation, the RWB may request that the Attorney General obtain a temporary restraining order and an injunction requiring compliance with all requirements. *Id.* § 13262. Similar relief is available when a person discharges waste or violates a WDR subject to the federal NPDES permitting program. *See generally id.* § 13385.

Porter-Cologne provides a number of enforcement mechanisms.

Common Water Quality Requirements for Mining

Surface mining typically involves a variety of discharges—from both point and non-point sources—that can adversely affect water quality. Specific WDR or NPDES requirements for a particular operation can only be determined case by case, depending on the type of mining, its location, the presence of adjacent water bodies, and the various standards imposed by the governing RWB and basin plan. Nonetheless, the following regulations commonly govern WDRs and NPDES permits for surface mining.

Additional Waste Discharge Reporting Obligations

Beyond generally-applicable reporting requirements, the Porter-Cologne Act mandates that a mining waste discharger submit two additional reports. A report covering "the physical and chemical characteristics of the waste that could affect its potential to cause pollution or contamination" must be submitted. Water Code § 13260(k)(1). A second report must review the possibility that the discharge might produce, over

Mining operators must provide two reports in addition to generally-applicable discharge reporting requirements.

the long term, acid mine drainage, the discharge or leaching of heavy metals, or the release of hazardous substances. *Id.* § 13260(k)(2).

Acid mine drainage, which can harm aquatic species, occurs when metallic mineral exposure to oxygen and water creates acidic discharges.

These reports enable the RWB to assess the potential for pollution from various types of metallic mining. Acid mine drainage, in particular, can be very harmful with effects that, if unaddressed, are long-lasting. Acid mine drainage occurs when mining exposes certain metallic substances in the earth's surface to oxygen and water, causing acidic discharges into groundwater and nearby streams, rivers, lakes, and other water bodies. The acid contamination lowers the pH, adversely affecting aquatic life and causing other significant environmental damage. Heap leaching and other forms of solution mining cause similar contamination (*see* chapter 1, describing solution or heap mining).

Storage of Overburden, Waste Rock, and Similar Mining Waste

The Porter-Cologne Act regulates the land disposal of mining wastes that can indirectly cause water degradation.

Regulation of waste disposals to land. Because the Porter-Cologne Act regulates non-point source discharges that can affect ground water, the discharges also include waste disposal to land. This can indirectly cause water degradation. Water in contact with mining waste piles containing contaminants, for example, can seep into groundwater and degrade water quality.

Mining waste is separated into three categories depending on the potential of the waste to degrade water quality.

Classification of mining waste. Specific requirements that apply to waste disposal relate to the severity of the potential to degrade water quality. **Group A**, with the highest potential, is considered hazardous waste. Cal. Code Regs. tit. 27, § 22480(b)(1). **Group B**, which may include hazardous waste, does not pose a significant risk to water quality. *Id.* § 22480(b)(2)(A). Group B also includes wastes that contain non-hazardous soluble pollutants of concentrations that either exceed water quality objectives or could degrade state waters. *Id.* § 22480(b)(2)(B). **Group C** complies with the governing water quality control plan. *Id.* § 22480(b)(3).

In determining whether to use a Group B or Group C classification, the RWB may consider whether the mining waste: (1) contains only low concentrations of hazardous constituents; (2) has no or low acid-generating potential; and (3) given its intrinsic properties, may be contained by less stringent measures. *Id.* § 22480(c)(1)–(3). Mining waste is also categorized as existing or new. An **existing unit** is one that existed prior to when the waste discharge to land regulations were adopted in 1984. *Id.* § 20080(d). A **new unit** is one constructed after 1984.

Standards for mining waste disposal. Overall, the regulations provide that mining waste "shall be treated or neutralized whenever feasible to minimize the threat to water quality and minimize the need to install waste containment structures." Cal. Code Regs. tit. 27, § 22480(d). In implementing this standard, the regulations address various aspects of waste disposal, including the following.

Mining waste must be located outside areas of rapid geologic change.

Siting. A new mining unit must be located outside an area of rapid geologic change. Cal. Code Regs. tit. 27, § 22490(a)(2). Only a Group C unit may be placed on

Holocene faults, and then only if displacement will not allow escape of wastes or cause irreparable damage to any containment structure. *Id.* § 22490(a)(1).

Flood protection. A new Group A waste unit may not be placed within the 100-year flood plain. Cal. Code Regs. tit. 27, § 22490, tab. 1.2. For existing Group A and new or existing Group B units, the discharger must protect the units from the 100-year peak stream flow. *Id.* A new Group C waste unit may be placed within the 100-year flood plain, but must preclude increased sediment in surface water. *Id.* Existing Group C units need only be retrofitted as needed to protect surface water quality. *Id.*

The mining waste regulations limit the locations of mining waste within 100-year flood plains.

Construction. Group A and Group B waste units are subject to specific construction standards. The very technical and specific regulation prescribes, for example, whether synthetic or clay liners must be placed beneath a containment unit and whether double liners are required, depending on unit type (*e.g.*, waste pile, surface impoundment, or tailings pond). *See* Cal. Code Regs. tit. 27, § 22490(f), tab. 1.3. Where waste will be placed in a containment structure, certain standards are prescribed. *Id.* § 22490(e); *see generally id.* § 20320.

Group A and B wastes are subject to specific construction standards.

Precipitation and drainage control. A mining discharger must design and construct a diversion facility to accommodate the anticipated precipitation volume. Cal. Code Regs. tit. 27, § 22490(h)(1). The facility must also accommodate peak flow from surface runoff. For a Group A waste unit, the facility must be able to accommodate a peak flow of one 25-year, 24-hour storm. *Id.* § 22490(h)(1)(A). Both Group B and C wastes must be able to handle a 10-year, 24-hour storm. *Id.* § 22490(h)(1)(B)–(C).

Mining operations must construct diversion facilities to accommodate precipitation.

Where a containment structure is not used to divert access runoff for Group A and B wastes, the runoff shall be collected and managed through a leachate collection and removal system. *Id.* § 22490(h)(2); *see id.* §§ 20340(b)–(e), 22490(g) (prescribing standards for leachate collection and removal). Additional requirements govern precipitation and drainage control. *See id.* § 22490(h)(3).

Monitoring. Mining discharges involving Group A and Group B wastes must comply with certain monitoring and corrective action requirements. Cal. Code Regs. tit. 27, § 22500(a). WDRs for a discharger of Group A and Group B waste require a detection monitoring program approved by the RWB. *Id.* § 20420. If leaks are detected, the discharger must further implement an evaluation monitoring program. Id. § 20425. If monitoring establishes that leaks have affected water quality, the discharger must take corrective action. *Id.* § 20430.

Mining operators that discharge Group A and Group B wastes must comply with monitoring and reporting requirements.

Closure. A mining waste unit must be closed as part of the reclamation operation. The standards governing closure of mining waste units are discussed in chapter 5.

Mining waste units must be closed as part of reclamation.

Dewatering and Process Water Disposal

When mining occurs in a quarry bowl or similar excavation, a process known as **dewatering** is often necessary. Because of the composition of the excavated deposits,

The disposal of water that accumulates in pits or excavations is a point source discharge subject to NPDES requirements.

water that is below the water table can seep into and inundate the bowl or excavation. Where this occurs, the water must be removed and transported elsewhere on the property to allow for mineral extraction. Dewatering generates wastewater and involves a point source discharge subject to federal NPDES requirements. *See, e.g.*, 40 C.F.R. § 436.31(b) (defining "de-watering" as "water that is impounded or that collects in the mine and is pumped, drained or otherwise removed from the mine through the efforts of the operator.")

Disposal water used for washing and processing is a point source discharge that must comply with the NPDES provisions of the CWA.

Another common mining discharge involves water used for a washing operation or another processing purpose. *See id.* § 436.31(e) (defining "process generated waste water" as "waste water used in the slurry transport of mined material, air emissions control, or processing exclusive of mining"). As with water removed through dewatering, water remaining after washing or processing must be collected and discharged somewhere onsite. Storage of this excess water is also subject to NPDES requirements.

Dewatering and process water discharges must usually be disposed in settling ponds and meet effluent limitations.

For water from either of these operations, a WDR typically requires that the discharged water be stored at a specified settling pond. WDRs may also prescribe the dimensions and locations of the pond, as well as the maximum volume of water they can contain.

Originating from a point source, dewatering and process water discharges are subject to effluent limitations under federal NPDES requirements. These usually regulate the pH of the discharge water and the settling pond, requiring both to fall within a pH range of 6.0 to 9.0. *See id.* § 436.32(a)(2) (imposing restriction for sand and gravel operations); *id.* § 436.22(a)(2) (imposing restriction for crushed stone operations). If the water contains chemical additives or other constituents—whether used as a part of the processing operations or occurring naturally in a rock deposit being mined—WDRs may impose additional effluent requirements that limit the chemicals and constituents to certain concentrations. *See generally* 40 C.F.R. Part 436 (prescribing effluent limitations for several types of mining operations). In addition, WDRs regulating dewatering and process water discharges often require periodic testing, monitoring, and reporting to ensure compliance with all effluent limitations.

Storm Water Runoff

The CWA contains special provisions that regulate storm water runoff.

The CWA contains special provisions that regulate storm water runoff from an industrial operation. *See generally* 33 U.S.C. § 1342(p)(3)(A). Nearly every mining operation must comply with these regulations. *See* 40 C.F.R. § 122.26(b)(14)(iii). An operation is exempt only if its storm water runoff is composed entirely of flows from conveyances or conveyance systems (*e.g.* pipes, conduits, ditches, or canals) that will not come into contact with any overburden, raw material, intermediate or finished product, or waste products. *Id.* § 122.26(a)(2).

While a discharger may obtain an individual permit that meets storm water discharge requirements, the EPA has approved a general California permit for industrial discharges, making compliance less burdensome and time-consuming than for the individual WDR process. An operator utilizing the general industrial permit must comply with applicable federal effluent limitations, and may not cause or contribute to the exceedance of any groundwater quality standards contained in the state water plan or applicable basin plan. State Water Res. Control Bd., *Water Quality Order No. 97-03-DWQ, Waste Discharge Requirements for Discharges of Storm Water Associated With Industrial Activities Excluding Construction Activities,* 3–4. Each facility utilizing the general permit must also develop a storm water pollution prevention plan (SWPPP). *Id.* p. 11.

The EPA has approved a general California permit for industrial stormwater discharge, which covers mining.

The California general industrial permit requires that a mine operator prepare a SWPPP.

The purpose of the SWPPP is to identify pollutants the operator may discharge that may affect the quality of stormwater discharges, and implement best management practices to reduce or prevent pollutants when storm water is discharged. *Id.* BMPs are divided into two types: **non-structural BMPs**, such as activity schedules, prohibitions of certain practices, maintenance procedures, and other low-cost measures; and **structural BMPs**, consisting of treatment measures, runoff controls, and overhead coverage. *Id.* An operator must also comply with monitoring and reporting requirements as part of the SWPPP. *Id.* pp. 24–40.

SWPPP = Storm water pollution prevention plan

in this chapter...

chapter 12

Wetlands, Streams, and Lakes

Surface mining not only affects air and water quality, but can disturb wetlands, streams, and lakes protected by federal and state law. An essential component of the ecosystem, wetlands play an important role in improving water quality, providing fish and wildlife habitat, and assisting in flood control. Since wetlands are such an essential component of the ecosystem, federal and state law accords them significant protection. Because the subject of such regulation is so broad—encompassing extensive federal and state regulations, case law, and other requirements—only a brief overview of the subject is provided here. For a more thorough understanding of wetlands regulation, readers should consult Paul D. Cylinder *et al.*, *Wetlands, Streams, and Other Waters: Regulation, Conservation, and Mitigation Planning* (Solano Press).

Wetlands Regulations Under Section 404 of the Clean Water Act

Applicability of Section 404

A mining operation that will disturb wetlands must obtain a **dredge or fill** permit mandated by Section 404 of the Clean Water Act (CWA). *See generally* 33 U.S.C. § 1344. Whether a Section 404 permit is necessary involves a two-step process. First, an operator must determine whether any part of the proposed mine site meets the following criteria for a wetland:

If a mining operation will disturb wetlands, the operator must obtain a Section 404 permit.

CWA = Clean Water Act

- Its soils are periodically or permanently saturated or inundated
- The soils are hydric, exhibiting anaerobic conditions in their upper parts
- The soils contain hydrophitic vegetation—plants that grow in soils where little oxygen is present

33 C.F.R. § 328.3(b)

If any part of the mine site constitutes a wetland under this test, the operator must next determine whether that part is also a "water of the United States" regulated by the CWA. *See id.* § 328.3(a) (defining "waters of the United States"). Whether such waters are present, in turn, depends on the ability of a particular wetland to affect interstate commerce. This is because the federal government's power to enforce Section 404 is limited to the extent of Congress' power to regulate interstate commerce under the U.S. constitution. *See* U.S. Const. art. I, § 8 cl. 3.

Whether small and isolated wetlands are within that power is a matter of considerable debate following the United States Supreme Court's decision in *Solid Waste Agency of N. Cook County v. United States Army Corps of Eng'rs* (2001) 531 U.S. 159 (*SWANCC*). Very recently, the issue was further muddied by the Supreme Court's decision in *Rapanos v. United States* (2006) 126 S.Ct. 2208, in which a fractured court failed to provide a majority opinion resolving the issue. As additional regulations and litigation are likely to follow from *Rapanos*, it is unfortunately not possible at the time of this writing to predict if, and to what extent, small and isolated wetlands will be subject to federal wetlands regulations.

Corps = U.S. Army Corps of Engineers

Wetlands Delineations

Wetlands delineations allow one to determine if jurisdictional wetlands exist on the property.

PRACTICE TIP

Because of potential penalties that exist for damaging wetlands, a mining applicant should conduct a wetlands delineation at a very early stage in the process.

Because of the uncertainty remaining after *SWANCC* and *Rapanos*, an applicant for a new mining operation should conduct a wetland delineation early in the process. A private consultant may be retained to determine whether any jurisdictional wetlands exist on the proposed mine property, according to the standards provided by the U.S. Army Corps of Engineers ("Corps"). *See generally* U.S. Army Corps of Eng'rs., *Minimum Standards for Acceptance of Preliminary Wetlands Delineation* (Nov. 30, 2001). The Corps then reviews the delineations. 33 C.F.R. § 325.9. Once issued, Corps wetlands delineations are valid for at least three years. U.S. Army Corps of Eng'rs, *Regulatory Guidance Letter 90-06* (Aug. 14, 1990).

Obtaining a wetlands delineation is critical. If no jurisdictional wetlands are present, an applicant is assured that the mining operation will not be subject to potential consequences for unlawful activity. *See, e.g.*, 33 U.S.C. § 1319 (providing for civil, criminal, and administrative penalties for unlawful dredging or filling of wetlands). By contrast, if jurisdictional wetlands do exist, the project can be reevaluated and revised to avoid disturbance, and the operator can avoid the burdens and expenses associated with a Section 404 permit.

Obtaining Section 404 Permits and Nationwide Permits

Mine operators that will dredge or fill wetlands must apply to the U.S. Army Corps of Engineers for 404 permits.

Where dredging or filling wetlands or other waters of the United States cannot be avoided, a mining operator must obtain a Section 404 permit from the Corps. Generally, this requires the applicant to show that no "practicable alternative" to the proposed discharge exists. 40 C.F.R. § 230.10. An alternative is "practicable" if it is "available and capable of being done after taking into consideration costs, existing technology, and logistics in light of overall project purposes." *Id.* § 230.3(q), 230.10(a)(2). Overall, this standard is difficult to meet.

DFG = Department of Fish and Game
NWP = Nationwide Permit Program

In some cases, the Nationwide Permit Program (NWP) can ease the burden of obtaining an individual permit from the Corps. Unlike a Section 404, the NWP acts as a general permit for certain classes of discharges and is much less demanding and easier to obtain. For certain mining activities, the Corps has adopted NWP 44. However, the San Francisco District of the Corps has rejected its use, making NWP 44 inapplicable to Bay Area and North Coast areas.

Mine operators may be able to avail themselves of NWP 44, which can ease the burden of obtaining a 404 permit.

Where applicable, NWP 44 authorizes discharges of dredged or fill material in certain waters where the mining and support areas do not have an impact on more than one-half acre. NWP 44, condition (a). Among other requirements, NWP 44 precludes mineral processing and beneficiation within 200 feet of the ordinary high-water mark of any open water body. *Id.* condition (h). Aggregate mining is also generally not permissible in streambeds or within 100 feet of the high-water mark of headwater stream segments where the annual average water flow exceeds one cubic foot per second. *Id.* condition (j). In addition, hard rock/mineral mining is prohibited in streams.

California Streambed and Lake Alteration Agreements

Under California law, an operator must notify the Department of Fish and Game (DFG) where mining would either (1) substantially divert or obstruct the natural flow of, or (2) substantially change or use the material from the bed, channel or bank of, any river, stream, or lake. Fish & Game Code § 1602(a). Notification is also required where an operator would dispose of debris, waste, or other material containing crumbled, flaked, or ground pavement into the water bodies. *Id.*

Mining operations that would interfere in certain ways with rivers, streams, or lakes must obtain DFG approval.

Upon receiving notification, DFG must determine if the mining impacts would substantially and adversely affect an existing fish or wildlife resource or otherwise affect a streambed. Its jurisdiction extends beyond the ordinary high-water mark regulated by the Corps to the highest flood flow mark and into adjacent riparian areas. If DFG determines that adverse impacts would exist, the operator must reach agreement with DFG regarding conditions to protect the fish and wildlife. *Id.* §§ 1602(a)(4)(B)–(C), 1603. Anyone who fails to provide the required notification or who violates any term of any agreement with DFG is subject to civil penalties of up to $25,000 for each violation, and may be enjoined from further violations. *Id.* § 1615(a), (e).

chapter 13

Endangered and Threatened Species

Both federal law and California law protect endangered and threatened plant and animal species from all types of human and industrial activity, including mining. Because endangered species laws and regulations are voluminous, complex, and ever-changing as a result of constant litigation, a comprehensive survey of those laws and regulations is beyond the scope of this book. Only a brief overview is therefore provided below. For more comprehensive coverage of the subject, readers should consult Kenneth M. Bogdan, *et al., Endangered Species Act, A Practical Guide to Compliance* (Solano Press, forthcoming).

The ESA prohibits the taking of endangered and threatened plant and animal species.

Federal Endangered Species Act

At the federal level, the Endangered Species Act (ESA) prohibits the taking of any endangered plant species on federal property and the taking of any endangered animal species anywhere. 16 U.S.C. § 1538(a). Similar restrictions protect species considered "threatened," though not endangered. *See generally* 50 C.F.R. § 17.71.

ESA = Endangered Species Act
USFWS = U.S. Fish and Wildlife Service

Listing of Endangered and Threatened Species

The designation of species protected by the federal ESA is accomplished through a listing process. After conducting a review of their status, the U.S. Fish and Wildlife Service (USFWS) identifies endangered and threatened species based on "the best scientific and commercial data available." 16 U.S.C. § 1533(b)(1)(A). Private individuals or entities, including environmental groups, may also propose adding a new species to the list.

Endangered and threatened species are identified through a listing process conducted by the USFWS.

USFWS applies a number of factors to determine if a species is threatened or endangered.

In reviewing its status, the USFWS may list a species as endangered or threatened if any of the following jeopardizes its continued existence:

- The present or threatened destruction, modification, or curtailment of its habitat or range
- Overutilization for commercial, recreational, scientific, or educational purposes
- Disease or predation
- The inadequacy of existing regulatory mechanisms
- Other natural or manmade factors affecting its continued existence

Id. § 1533(a)(1)

Based on these factors, an "endangered" species is "any species which is in danger of extinction throughout all or a significant portion of its range...." *Id.* § 1532(6). A "threatened" species is any species that "is likely to become an endangered species within the foreseeable future throughout all or a significant portion of its range." *Id.* § 1532(20).

As part of the listing process, USFWS designates the endangered or threatened species' critical habitat.

As part of the listing process, the USFWS must designate "critical habitat" for endangered or threatened species. This includes:

- Specific areas within the geographical area currently occupied by a species... on which are found those physical or biological features (i) essential to the conservation of the species and (ii) that may require special management consideration or protections
- Specific areas outside the geographical area occupied by a species at the time it is listed upon a determination by the Secretary that such areas are essential for the conservation of the species

50 C.F.R. § 424.02(d)

Taking Prohibitions and Incidental Take Permits

A taking occurs under the ESA by harming a species or significantly modifying its habitat.

Taking an endangered or threatened species under the ESA means to "harass, harm, pursue, hunt, shoot, wound, kill, trap, capture, or collect, or to attempt to engage in any such conduct." 16 U.S.C. § 1532(19). A taking also occurs when harm causes "significant habitat modification." 50 C.F.R. § 17.3; *see Babbitt v. Sweet Home Chapter of Com'tys for a Greater Oregon* (1995) 515 U.S. 687, 708 (upholding the extension of a taking under the ESA to habitat modification).

The ESA has been amended to allow the incidental taking of a species so long as a person or entity has submitted a HCP.

Since its initial adoption, the ESA has been amended to permit an "incidental" taking of endangered and threatened species. To protect against an inadvertent taking, a person or entity may obtain an **incidental take permit.** 16 U.S.C. § 1539(a)(1)(B). To obtain the permit, one must submit a habitat conservation plan (HCP) describing:

HCP = Habitat conservation plan

- Impacts that are likely to result from a taking
- Steps the applicant will take to mitigate and minimize the impacts
- Funding available to implement the steps
- Alternative actions to the taking the applicant has considered

- The reason why alternatives are not being adopted
- Other measures the [USFWS] may require

Id. § 1539(a)(2)(A)

Particularly in California, HCPs are becoming an important feature of ESA compliance. Penalties for violating the ESA can be harsh. *See id.* § 1540(a)(1) (providing for fines of up to $25,000 for each violation). HCPs allow a person or entity to avoid the consequences of unintentionally violating the ESA.

CESA = California Endangered Species Act

California Endangered Species Act

The state also protects endangered and threatened species through the California Endangered Species Act (CESA). *See generally* Fish & Game Code § 2050 *et seq.* As with the federal ESA, CESA establishes a list of endangered and threatened species. *Id.* § 2070. While also prohibiting the taking of endangered species, CESA, unlike federal law, applies equally to wildlife and plant species, regardless of where the species are found. *Id.* § 2080.

CESA prohibits the taking of wildlife and plant species, regardless of where the taking occurs.

Like the federal ESA, CESA further provides for an incidental take permit. *Id.* § 2081(b). Among other requirements, the permit may be issued when:

Like ESA, CESA allows a person or entity to obtain an incidental take permit.

- Impacts of the authorized take are minimized and mitigated
- Measures authorized as part of the take are roughly proportional to the impact on a species
- The take will not jeopardize the continued existence of the species
- Applicant ensures adequate funding to meet all implementation measures

Id. § 2081(b)–(c)

chapter 14

Handling of Hazardous Materials

Surface mining may require the use chemicals as part of its operation. Under California law, some chemicals may be considered hazardous materials, requiring special handling and reporting. Where these materials are used, an operator must prepare a hazardous materials business plan, which is required to ensure sound emergency planning and allow for prompt and effective emergency response. *See* Health & Safety Code § 25504.

When using chemicals that are hazardous materials, mining operators must prepare hazardous materials business plans.

The business plan requirement applies to any person who handles a minimum quantity of any "hazardous material," broadly defined to include:

> [A]ny material that, because of its quantity, concentration, or physical or chemical characteristics, poses a significant present or potential hazard to human health and safety or to the environment if released into the workplace or the environment. 'Hazardous materials' include, but are not limited to, hazardous substances, hazardous waste, and any material that a handler or administering agency has a reasonable basis for believing that it would be injurious to the health and safety of persons or harmful to the environment if released into the workplace or the environment.
>
> *Id.* § 25501(o)

To trigger the business plan requirement, an operator must handle—at any one time during a year—an amount that is equal to, or greater than, 500 pounds or 55 gallons, or 200 cubic feet if the hazardous material is compressed gas. *Id.* § 25503.5(a)(1)(A)–(B).

Operations must handle certain quantities of hazardous materials to be subject to the business plan requirement.

An operation subject to the business plan requirement must obtain approval from a governing local agency known as a **certified unified program agency** or **unified program agency**. *See generally id.* §§ 25504, 25505. The principal component of the business plan is the hazardous materials inventory. *Id.* § 25504(a). For every hazardous

Business plans must provide an inventory of all the hazardous materials the operation handles.

substance handled, the inventory must disclose its chemical name and common name, the category of waste for every hazardous waste, and the maximum amount of hazardous material and waste handled at any one time during the year. *Id.* § 25509(a)(1)–(4). It must also indicate the onsite location of every identified material and waste to ensure that fire, safety, and health personnel can utilize that information in an emergency. *Id.* § 25509(a)(5).

Business plans must provide an emergency response and evacuation plan, describe employee training measures, and identify measures to mitigate any release of hazardous material.

In addition, a business plan must:

- Provide an emergency response plan and procedures for releases or threatened releases of hazardous materials
- Discuss procedures to mitigate any release or threatened release to minimize harm to persons, property, or the environment
- Create evacuation plans and procedures
- Provide for employee training and safety procedures and for handling of releases or threatened releases of hazardous materials

Id. § 25504(b), (c)

Whenever the release or threatened release of hazardous materials is discovered, the operator must report that release to its local governing agency and to the office of emergency services. *Id.* § 25507(a). Anyone who violates these provisions may be imprisoned, fined, or subject to other significant penalties. *See id.* §§ 25514–25515.

Glossary

404 Permit
A common reference to **dredge or fill permits** under **Section 404** of the **Clean Water Act**.

A.B. 3098 List
A list prepared quarterly by the **Office of Mine Reclamation** of operators eligible to sell material to state and local agencies. To be included on the list, operators must have an approved **reclamation plan** and **financial assurance** or must have a **reclamation plan appeal** or **financial assurance appeal** pending.

Abandoned mine
A **mining operation** that concluded without undergoing **reclamation**.

Abandonment of nonconforming use
The cessation of a **nonconforming use** for more than the period prescribed by the applicable **zoning ordinance** (*e.g.,* 180 days, one year), resulting in the termination of the use's right to continue operating as a lawful nonconforming use.

Acid mine drainage
Acidic discharges that result when mining exposes certain metallic substances in the earth's surface to oxygen and water. These discharges can seep into groundwater and nearby streams, rivers, lakes, and other water bodies, adversely affecting aquatic life and damaging the environment.

Administrative mandamus
A type of lawsuit that challenges a local agency's discretionary decision following a public hearing. It is usually the only method in which aggrieved parties may challenge city and county land use decisions. *See* Cal. Pub. Res. Code §§ 1094.5, 1094.6.

Administrative proceedings
Under **SMARA**, a special procedure a **lead agency** or the **Department of Conservation** may follow to enforce the Act against non-compliant operators. After providing the **operator** notice of its violations, the lead agency or Department issues an order to comply and holds a public hearing. If the operator does not comply with the order following the hearing, it is subject to fines up to $5,000 per day. *See* Cal. Pub. Res. Code §§ 2774.1, 2774.2.

Aggregate
A term used to refer to sand, gravel, and rock used for road, highway, and building construction.

Aggregate Resource Area
An additional **classification** given to areas classified as **MRZ-2a** or **MRZ-2b** if they contain **aggregate** deposits. Depending on their degrees of significance, these areas are divided into immediately significant, highly significant, and significant areas.

Air basin
An area of the state that is either an **attainment** or **non-attainment** area under the **Clean Air Act** and California's air quality statutes and regulations.

Amortization
The restriction of a **nonconforming use** to a defined term after which the use must comply with **zoning**. The amortization period must provide the owner of the use sufficient time to recoup its investment.

Annual inspection
Under **SMARA**, an annual inspection the **lead agency** must perform of every **surface mining operation** within its jurisdiction. The purpose of the inspection is to determine the operator's compliance with its reclamation plan and SMARA.

Annual report
Under **SMARA**, a report **operators** must annually submit to the Department of Conservation describing, among other things, the production, acreage disturbed, and acreage reclaimed during the previous year.

Anti-degradation policy
A policy states must adopt under the **Clean Water Act** to provide for a level of water quality necessary to ensure existing uses are maintained and protected. 40 C.F.R. § 131.12(a)(1).

Army Corps of Engineers
The federal agency responsible for issuing **dredge or fill permits** under **Section 404** of the **Clean Water Act**.

Asbestos
A naturally occurring fiber in serpentine rock deposits that can become airborne when freed by mining. Scientific studies have shown that prolonged exposure to asbestos fibers can lead to lung cancer, mesothelioma, and asbestosis.

Attainment area
Under the **Clean Air Act**, an area that meets the standard for a particular **NAAQS** pollutant.

Augur mining
The method of surface mining in which minerals are removed by boring from the **highwalls** of mined-out strips with special augur machines.

Authority to construct
An authorization usually required by air pollution control districts that regulates construction of machines, equipment, and facilities. The purpose of the authorization is to ensure compliance with the **Clean Air Act**, state air quality laws, and any regional district regulations.

BACT
Best available control technology

Baghouse
An air pollution abatement device attached to a mining plant that uses a vacuum-type cleaner to trap large particles.

Glossary

Base rock

A type of **aggregate** consisting of crushed rock that is placed beneath the surface of roads and highways.

Baseline

Under CEQA, a term used to describe the existing physical and environmental conditions associated with a **project**. Whether the project has the potential to create any significant adverse environmental impacts, requiring an **environmental impact report**, is determined by measuring the project's potential impacts against its baseline.

Basin plan

A common name given to **regional water quality control plans**.

BAT

Best available technology

Batch plant

A plant used to make concrete.

Benching

The downward, layered progression of excavation in the **open-pit method of mining**. Each layer is lower and smaller in area than the previous one, creating a step-like appearance. Benching is necessary to maintain safe slopes and to allow equipment access to and from the bottom of the excavation, or **pit**.

Beneficial use

A type of use of water, including domestic, municipal, agricultural, industrial, power generation, recreation, navigation, or preservation of fish, wildlife, and aquatic resources.

Best available control technology

A standard under the **Clean Air Act** that requires sources of air pollutants to achieve an emission level that, "taking into account energy, environmental, and economical impacts and other costs," is "achievable for such facility through application of production processes and available methods, systems, and techniques. . . ." 42 U.S.C. § 7475(a)(4), 7479 (3).

Best available technology

A standard under the **Clean Water Act** that corresponds to demonstrated technology that is considered economically feasible for a particular class of discharge.

Best management practices

Practices for managing waste discharges identified by a **regional water quality control board**. These practices are often identified in **regional water quality control plans** and furnish dischargers with a means for complying with their **waste discharge requirements**.

Blasting

In mining operations, the use of explosives to free rock masses from a mineral deposit.

BMPs

Best management practices

Borrow pit

A pit or **working** area excavated for purpose of providing material to fill another site or for use in a large project.

Bowl

A term often used to describe the excavation, or **working area**, of a **quarry** where rock, stone, or gravel are mined.

CAA

Clean Air Act

California Air Resources Board

An agency of the State of California that has adopted California's air quality standards and California's **state implementation plan** for purpose of complying with the **Clean Air Act**.

California Department of Fish and Game

The state agency responsible for issuing **streambed alteration agreements**.

California Environmental Quality Act

Known more commonly by its acronym, CEQA, a comprehensive statute requiring public agencies to evaluate the environmental impacts of every discretionary public **project** they undertake or private project for which their authorization is sought. CEQA has particular application to mining because of its land-altering nature and potential for adverse environmental impacts. Typically, cities and counties do not decide whether to approve mining projects unless they have prepared a full **environmental impact report**.

California Geological Survey

An agency within the **Department of Conservation** consisting of geologists who, among other duties, conduct an ongoing inventory of the state's mineral resources for purposes of SMARA's **classification** provisions.

California Vehicle Code

A set of laws adopted by the legislature to regulate vehicle traffic throughout the state. In addition to governing the rules of the road (*e.g.,* speeding, traffic safety), this code regulates heavy trucks in terms of their maximum allowable **load** weights and the abilities of cities and counties to restrict the roads and highways on which they may travel.

California Water Plan

A comprehensive plan the **State Water Resources Control Board** has adopted to guide the protection, conservation, development, and utilization of the state's water resources.

CARB

California Air Resources Board

Categorical Exemption

An activity that is considered exempt from CEQA as set forth in the **CEQA Guidelines**.

CEQA

California Environmental Quality Act

CEQA Guidelines

The administrative regulations adopted to interpret and clarify **CEQA**, appearing in California Code of Regulations, Title 14, §§ 15000–15387.

CGS

California Geological Survey

Circulation

Following completion of a **draft environmental impact report**, the release of the report for mandatory public review and comment.

Classification

In mining operations, the separation of processed material into desired classifications of salable products through screening equipment.

Classification of mineral lands

See **mineral land classification**.

Glossary

Clean Air Act
A comprehensive federal act that establishes specific air quality standards, requires states to adopt plans to achieve those standards, and specifies criteria that certain new, major emitting air pollution sources must meet. *See generally* 42 U.S.C. § 7401 *et seq.*

Clean Water Act
A comprehensive federal law that regulates water quality. The law regulates mining through the **National Pollutant Discharge Elimination System** and by controlling disposals of dredged or fill material into **navigable waters**.

Comminution
Following **fragmentation** in mining operations, the progressive reduction of excavated material to smaller, specified sizes. This is accomplished through special **crushing** equipment.

Conditional use permit
A specific authorization by a city or county to carry on a land use at a particular location. **Zoning ordinances** allow these uses in the applicable zoning district, but make them conditional to ensure that potential land use incompatibilities are addressed before the uses are permitted. Authorization for the use must occur following notice to affected landowners and a public hearing.

Consistency doctrine
The doctrine requiring that all land use decisions and enactments be consistent with the policies and goals of a city or county's **general plan**. Failure to achieve consistency could result in the enactment or approval being overturned if challenged in court.

Critical habitat
Under the **Endangered Species Act**, certain areas the **United States Fish and Wildlife Service** designates as essential to the conservation of **endangered** and **threatened species**.

Crushing
In mining operations, the progressive reduction of excavated material through special crushing equipment.

CUP
Conditional Use Permit.

Cyclone
An air pollution abatement device attached to a mining plant that uses centrifugal force to remove particles from polluted air.

Department of Conservation
The umbrella state agency responsible for enforcement and interpretation of SMARA. Within the Department are the **State Geologist**, **California Geological Survey**, **State Mining and Geology Board**, and **Office of Mine Reclamation**, which each perform distinct functions under the Act.

Designation
Under **SMARA**'s mineral conservation provisions, if an area has been classified **MRZ-2a** or **MRZ-2b**, the **State Mining and Geology Board** may designate the area as a significant statewide or regional mineral resource. Such designations trigger a **lead agency** responsibility to consider the importance of the affected mineral resource and balance mineral values against alternative land uses when considering any mining application.

Designation appeal
An appeal to the **State Mining and Geology Board** of any **lead agency** decision approving or denying a mining permit application in an area designated as a **significant mineral resource**. If the Board rules in the appellant's favor, the Board remands to the lead agency to reconsider its decision.

Development
The second phase of a surface mine in which the mining operator makes necessary property and business arrangements and secures all government authorizations needed to permit a mine site.

Dewatering
The removal of water from **pits** or **working areas** to allow for mineral extraction.

DFG
California Department of Fish and Game.

Diminishing asset doctrine
A special doctrine applicable to extractive **nonconforming uses**. Because these uses do not operate within fixed locations, the doctrine provides that such uses lawfully include the entire area the operator intended to mine, as demonstrated by objective evidence, at the time the applicable zoning ordinance was adopted. *See Hansen Brothers Enters., Inc. v. Bd. of Supervisors*, 12 Cal. 4th 533 (1996).

Draft environmental impact report
The completed environmental impact report, referred to as a "draft" because it has not yet undergone **circulation**.

Dredge or fill permit
A permit that must be obtained from the **United States Army Corps of Engineers** for the discharge of dredged or filled material into **waters of the United States**, including **jurisdictional wetlands**. To obtain approval, an applicant must generally show that there is no "practical alternative" to the proposed discharge. *See* 40 C.F.R. § 230.10.

Dredging
A type of **surface mining** involving the underwater extraction of a mineral deposit from a special floating machine known as dredge.

Effluent limitations
Under the **Clean Water Act**, restrictions on the "quantities, rates, and concentrations of chemical, physical, biological, and other constituents which are discharged" into regulated waters. 33 U.S.C. § 1311.

EIR
Environmental impact report.

End use
The post-mining land use to which a mine site will be reclaimed following conclusion of mining. Also referred to as a **second use**.

Endangered species
Under the **Endangered Species Act**, "any species which is in danger of extinction throughout all or a significant portion of its range. . . ." 16 U.S.C. § 1532(6).

Endangered Species Act
A comprehensive federal law that prohibits the taking of any **endangered** or **threatened species**.

Environmental Impact Report
Under **CEQA**, a comprehensive document, usually referred to as an EIR, that analyzes a project's environmental impacts. The EIR must fully describe the project and its existing physical and environmental setting, analyze its potential environmental impacts, propose measures to mitigate such impacts, and describe alternatives to the project. Before certifying an EIR, the agency must circulate it for public review and comment.

Glossary

Environmental Protection Agency

A federal agency with several responsibilities for enforcing federal environmental laws, including the **Clean Water Act** and **Clean Air Act**.

Environmental setting

Under **CEQA**, a description of a project's existing physical and environmental characteristics, often referred to as the project's **baseline**.

EPA

Environmental Protection Agency

ESA

Endangered Species Act

Expansion (of a nonconforming use)

The unlawful operation of a **nonconforming use** beyond its permitted scope. The prohibited expansion may occur by adding new activities to mining operations that did not exist at the time the applicable zoning ordinance was adopted, installing plants or equipment that significantly expand production, or increasing production levels substantially.

Exploitation

The third phase of a surface mine in which production of the mineral commodity occurs.

Exploration

Following **prospecting**, a more refined evaluation of a specific site for its resources and reserves. Together with prospecting, exploration is considered the first phase of a surface mine.

Fair argument (standard)

The standard courts have established for determining when agencies must prepare **environmental impact reports** under **CEQA**. Whenever it can be fairly argued based on substantial evidence in the record that a project may have a significant environmental impact, an agency may not prepare a **negative declaration**. *See Friends of "B" Street v. City of Hayward*, 106 Cal. App. 3d 988, 1001 (1980).

Final environmental impact report

The completed environmental impact report following circulation. Unlike the **draft environmental impact report**, this version contains the **responses to comments** received from the public and interested agencies.

Financial assurance

A bond, letter of credit, trust fund, or other suitable financial instrument operators must post under SMARA to guarantee their performance of **reclamation** in accordance with their **reclamation plans**. *See* Cal. Pub. Res. Code § 2773.1.

Financial assurance appeal

An appeal by an operator to the **State Mining and Geology Board** following the lead agency's denial of financial assurances, denial of due process, or failure to timely process a financial assurance application. If the Board rules in favor of the operator, it must approve the financial assurances.

Financial assurance cost estimate

An estimate an **operator** must submit each year to its **lead agency** to ensure that the amount of its **financial assurance** remains sufficient to accomplish **reclamation**.

Financially incapable of reclaiming

Under **SMARA**, the status an **operator** attains when it (1) fails to provide the **lead agency** with a **financial assurance cost estimate** within 30 days of being requested to do so; (2) does not replace a lapsed **financial assurance** with a new one within 30 days; (3) is unable to be contacted by the lead agency after 90 days of becoming **idle**; or (4) remains idle for more than one year without obtaining approval of an **interim management plan**. *See* Cal. Code Regs., tit. 14, § 3811.

Findings

Written statements the lead agency prepares to describe the legal and factual bases for its decision to approve or deny a **project**.

Forfeiture of financial assurance

Under **SMARA**, the action the **lead agency** must take whenever an **operator** becomes **financially incapable of reclaiming**. Following prescribed notice and hearing requirements, the lead agency must forfeit the operator's **financial assurance** and use its proceeds to effect **reclamation** of the mine site.

Fragmentation

The process by which mining operators separate rock masses from a mineral deposit. This is accomplished by a number of means, including the use of ripping blades, dozers, and other forms of mechanical extraction. It can also be accomplished through the use of explosives, or **blasting**.

Fuel minerals

The class of **minerals** consisting of oil and gas.

General plan

A document every city or county must adopt to serve as the charter or blueprint for growth and development within its jurisdiction. The plan must be comprehensive and focused on the long term. It must include a number of elements, including the **Land Use Element**, that sets broad land use policies and specifies the geographic distribution of land uses, and the **Conservation Element** that establishes policies for protecting natural resources, including mineral resources.

Habitat conservation plan

A plan applicants may submit to the **United States Fish and Wildlife Service** as part of an **incidental take permit** application to mitigate and minimize the impacts of their proposed activities on **endangered** or **threatened species**.

Haul route

A designated route trucks must travel between a mine site and major thoroughfares and highways. The purpose of a haul route is to keep mining trucks off streets that are incompatible with heavy truck traffic.

Hauling

The transportation of **minerals** by truck from the mining operation to job site.

Hazardous material

Under state law, "any material that, because of its quantity, concentration, or physical or chemical characteristics, poses a significant present or potential hazard to human health and safety or to the environment if released into the workplace or environment. . . ." Cal. Health & Safety Code § 25501(o).

Hazardous material business plan

Under state law, plans that handlers of hazardous materials must prepare to ensure sound emergency planning and allow for prompt and effective emergency response. *See* Cal. Health & Safety Code § 25504.

Heap mining

A term often used to refer to **solution mining**.

Glossary

Highwalls

The walls remaining after a strip is removed by **open-cast mining**. These are often excavated with special machines as part of the **augur mining** method.

Hot plant

A plant used to make asphalt.

Hours of operation

The daily hours in which mining operations may be conducted, as established by the **lead agency**. Hours of operation are a common requirement of a **conditional use permit**.

Idle

Under **SMARA**, the status of a mining operation that curtails operations by more than 90 percent of its previous maximum annual mineral production.

Incidental take permit

An authorization the **United States Fish and Wildlife Service** may issue under the **Endangered Species Act** to permit the incidental taking of **endangered** or **threatened species**.

Industrial minerals

A reference sometimes used to describe **non-metallic minerals**.

Initial study

Under **CEQA**, a study agencies conduct to determine a project's potential environmental impacts. Typically, the agencies use a checklist contained in Appendix G of the **CEQA Guidelines**. The study is circulated to several departments within the agency as well as to other public agencies for purpose of determining a project's potential environmental effects. After being prepared, the initial study is used to determine whether the agency will require an **environmental impact report** or a **negative declaration**.

Interim management plan

Under **SMARA**, a plan **operators** must submit to their **lead agencies** whenever mining operations become **idle**. Operators who fail to submit interim management plans can be determined to have abandoned their operations.

Jurisdictional wetland

A term used to refer to wetlands that are regulated under **Section 404** of the **Clean Water Act**.

LAER

Lowest available emission rate

Lead agency

Under **SMARA**, the city or county in which a mine site is located. Lead agencies have several responsibilities for enforcing SMARA, as described in chapters 5 and 6.

Listing (of endangered or threatened species)

Under the **Endangered Species Act**, the identification of **endangered** or **threatened** species, conducted by the **United States Fish and Wildlife Service**.

Load

One truck load of mined material, typically ranging from 10 to 25 tons, depending on the size of trailer used.

Lowest available emission rate

A standard under the **Clean Air Act** that is the more demanding of: (1) the most stringent limitation contained in the **state implementation plan** for a class or category of sources; or (2) the most stringent limitation achieved in practice by the class or category or source.

Major emitting facility

Under the **Clean Air Act**, a source of pollutants in an attainment area that is subject to the Act's **prevention of significant deterioration** provisions. Mining uses subject to these provisions include Portland cement plants or phosphate rock plants that produce 100 tons or more of any **NAAQS** pollutant.

Major stationary source

Under the **Clean Air Act**, a building, structure, or facility that emits an air pollutant at a volume depending on the degree of an area's non-attainment status. These sources are subject to **New Source Review**.

Metallic minerals

The class of minerals consisting of ferrous metals (*e.g.,* iron and tungsten), base metals (*e.g.,* copper, lead, and zinc), and precious metals (*e.g.,* gold, silver, and platinum).

Mine

The site where **minerals** are excavated from the earth.

Mined lands

Under **SMARA**, "the surface, subsurface, and ground water of an area in which surface mining operations will be, are being, or have been conducted, including private ways and roads appurtenant to any such area, land excavations, workings, mining waste, and areas in which structures, facilities, equipment, machines, tools, or other materials or property which result from, or are used in, surface mining operations are located." Cal. Pub. Res. Code § 2729.

Mineral

An inorganic element or compound formed by many years of geologic processes. Minerals are divided into three classes: **metallic minerals**, **non-metallic minerals**, and **fuel minerals**.

Mineral extractive (or mineral resource) district

A type of **zoning district** some cities and counties use to zone mining uses. Sometimes these zoning districts are created as **overlay districts**. They typically establish the findings cities and counties must make before allowing mining and set minimum standards for protecting neighboring uses and the environment.

Mineral land classification

Required by SMARA, a comprehensive survey by the **California Geological Survey** of the state's mineral resources. Because of the enormous size of the California's mineral resources, classification is an ongoing process that, to date, has focused on the fastest-urbanizing areas of the state. Classification entails a geologic and economic evaluation of selected survey areas. The results are reported in **mineral land classification reports**, the information from which must be incorporated into **lead agency general plans**.

Mineral land classification report

A report of the **California Geological Survey**'s classification of lands within a selected survey area. Lands within the survey area are classified based on the presence and economic viability of mineral deposits. **Lead agencies** must incorporate the information from these reports in their **general plans**.

Mining

The extraction of any naturally occurring minerals for utilitarian purposes.

Mining wastes

Under the **Porter-Cologne Water Quality Control Act**, overburden, unprocessed excavated material, and sludges subject to the Act's **waste discharge to land** requirements.

Glossary

Mitigated negative declaration

Under **CEQA**, a **negative declaration** in which some aspects of the project may create significant adverse environmental impacts, but those impacts can be rendered insignificant through mitigation measures.

Mitigation measure

A provision in an **environmental impact report** or **mitigated negative declaration** proposing a restriction or condition designed to reduce one of the project's environmental impacts to a level of less than significant.

Model Surface Mining Ordinance

A model ordinance the **State Mining and Geology Board** has created to assist cities and counties in complying with **SMARA**'s requirement they adopt a **surface mining ordinance**. The ordinance can be found on the **Department of Conservation** website (*www.consrv.ca.gov*).

Modernization (of a nonconforming use)

The equipment of a mining use with modern plants and equipment in a way that does not constitute an unlawful **expansion of a nonconforming use**. Generally, additions to an operation do not constitute prohibited expansions if they simply improve on existing facilities and do not create substantially different and adverse impacts on surrounding land uses.

MRZ-1

Under SMARA's mineral conservation provisions, this classification is given if adequate geologic information rules out the presence of significant mineral deposits.

MRZ-2a

The highest classification given to a site, indicating the presence of an existing mine or an area where extensive sampling indicates the presence of a significant mineral deposit.

MRZ-2b

The second-highest classification given to a site, indicating an area where the significance of the resource is inferred or where the significance is demonstrated but does not meet economic criteria.

MRZ-3a

Areas given this classification are believed to have a moderate potential for discovery of economic mineral deposits.

MRZ-3b

Areas are given this classification if it is plausible that economic mineral deposits may be present.

MRZ-4

Areas classified MRZ-4 are those where there is a lack of knowledge regarding mineral occurrence.

NAAQS

National Ambient Air Quality Standards

National Ambient Air Quality Standards

Standards the **Environmental Protection Agency** has promulgated under the **Clean Air Act** to regulate air pollutants that can be reasonably anticipated to endanger public health and welfare.

National Pollutant Discharge Elimination System

A permitting scheme under the **Clean Water Act** that prohibits any person from disposing a pollutant from a point source into navigable waters

Navigable waters

Under the **Clean Water Act**, jurisdictional waters that broadly include **waters of the United States**. *See generally* 33 C.F.R. § 328.3(a).

Negative declaration

Under **CEQA**, an environmental document agencies prepare when they have determined that a **project** does not have the potential to create any significant adverse environmental impact. Agencies may only prepare a negative declaration when they determine there is no **fair argument** that such an impact would be created.

New source review

Under the **Clean Air Act**, a permitting scheme in which **major stationary sources** must achieve the **lowest available emission rate** and meet **offset** requirements.

NIMBY

A pejorative term that refers to the "not in my backyard" syndrome, in which residents are perceived as opposing a proposed land use for selfish reasons.

Non-attainment area

Under the **Clean Air Act**, an areas that does not comply with the standard for a **NAAQS** pollutant.

Nonconforming use

A land use that is not required to comply with **zoning** because it existed before adoption of the applicable **zoning ordinance**. The use is permitted to continue to avoid the unconstitutional taking of property that would occur if it were forced to immediately comply with zoning. However, the use must operate similarly to the way it did at the time zoning was first adopted and may not be discontinued for more than certain, prescribed periods (*e.g.*, 180 days or one year).

Non-metallic minerals

The class of minerals consisting of substances other than metallic minerals, such as sand, gravel, limestone, and sulfur.

NPDES

National Pollutant Discharge Elimination System

NSR

New source review

Office of Mine Reclamation

The state agency within the **Department of Conservation** responsible for assisting and overseeing **lead agency** enforcement of **SMARA** and monitoring mining **operator** compliance under the Act.

Offset

Under the **New Source Review** provision of the **Clean Air Act**, the requirement that a source of an air pollutant not increase the overall emission in a non-attainment region beyond that generated at the time of permit application by existing sources and proposed and modified non-major sources.

OMR

Office of Mine Reclamation

Open-cast mining

Sometimes referred to as **strip mining**, this is a method of **surface mining** in which excavation occurs laterally in successive strips. As one strip is mined out, it is filled with **overburden** from the next strip.

Open-pit mining

The most common type of **surface mining**. Minerals are removed in layers, with each successive layer being lower and smaller in area than the previous one. This downward, layered manner of excavation is known as **benching**, and often gives the excavation a bowl-like appearance.

Glossary

Operator

Under **SMARA**, "any person who is engaged in surface mining operations, himself, or who contracts with others to conduct operations on his behalf, except a person who is engaged in surface mining operations as an employee with wages as his sole compensation." Cal. Pub. Res. Code § 2731.

Ordinance

A legislative act of a city or county enacting a generally applicable law within its municipal limits. In the context of surface mining, operations are usually subject to ordinances regulating **zoning**. **SMARA** also requires every city and county to adopt a **surface mining ordinance** whenever any **operation** is located within their jurisdiction.

Overburden

Surface layers of uneconomic or marginally economic material that must be removed to expose the underlying **mineral** deposits.

Overlay district

A type of **zoning district** that is combined with another district (often referred to as the "base" district). The overlay district specifies additional regulations that apply for specific uses permitted by the base district. For instance, a number of cities and counties have created **mineral extractive districts** as overlays.

Permit Appeal

See **designation** appeal.

Permit to operate

An authorization usually required by air pollution control districts that regulates air pollution from the operation of machines, equipment, and facilities. The purpose of the authorization is to ensure compliance with the **Clean Air Act**, state air quality laws, and any regional district regulations.

Pit

A term typically used to refer to the excavations, or **working areas**, in **open-pit mining** operations.

PM_{10}

A pollutant regulated under the federal **National Ambient Air Quality Standards** and California's ambient air quality standards. PM10 consists of fine, breathable particles, the prolonged exposure to which can lead to aggravated asthma, chronic bronchitis, and decreased lung function.

Point source

Under the **Clean Water Act**, "any discernible, confined and discrete conveyance... from which pollutants are or may be discharged." 33 U.S.C. § 1362(14).

Porter-Cologne Water Quality Control Act

A comprehensive California law that regulates water quality. A key component of this law is a permitting scheme in which dischargers must obtain approval of **waste discharge requirements** from a **regional water quality control board**.

Portland cement

The type of cement used to make white paved surfaces.

Prevention of significant deterioration

Under the **Clean Air Act**, a permitting scheme for certain **major emitting facilities** in **attainment areas**. If applicable, these provisions require the sources to achieve **best available control technology**.

Project

Under **CEQA**, the whole of an action that has the potential for resulting in either a direct physical or reasonably foreseeable indirect change in the environment. *See* Cal. Pub. Res. Code § 21065; Cal Code Regs. tit. 14, § 15378(a).

Project alternatives

Under **CEQA**, a mandatory description in an **environmental impact report** of reasonable alternatives to the project or the project location. The alternatives discussion should describe a range of options that could avoid or substantially lessen the project's significant environmental effects while still accomplishing most of the project's objectives.

Project description

Under CEQA, a summary of the project for which environmental review is required. Among other things, the project description must identify the specific activities for which approvals are being sought, the project objectives, the affected lands, the duration of the proposed activities, anticipated production levels, and locations for ingress and egress.

Prospecting

The search for potential mine sites through review of geologic maps, reports and literature; review of aerial photography; and other methods of evaluation. Prospecting is often grouped with **exploration** as the first phase of a surface mine.

PSD

Prevention of significant deterioration

Quarry

A term typically used to refer to **open-pit mining** operations where rock, stone, or gravel is mined.

Reclamation

The comprehensive process in which mined property is converted to a post-mining land use, or **end use**. Reclamation typically involves removal of mine facilities and equipment, reshaping of the mined land, **revegetation**, and remediation of any adverse environmental conditions—all in a manner such that the end use can be implemented. Under **SMARA**, every **operator** of a **surface mining operation** must obtain approval of a **reclamation plan** and post a **financial assurance** to guarantee it will reclaim its mining site. *See* Cal. Pub. Res. Code § 2733 (defining "reclamation" under SMARA).

Reclamation appeal

An appeal by an **operator** to the **State Mining and Geology Board** following the **lead agency**'s denial of a **reclamation plan**, denial of due process, or failure to timely process the plan. If the Board rules in favor of the operator, it must approve the reclamation plan.

Reclamation bond

The most commonly used form of **financial assurance** in which a third party, usually a bank or surety company, guarantees to pay the amount of the assurance should the **operator** become **financially incapable of reclaiming**. To secure the bond, the operator ordinarily must pay annual premiums.

Reclamation plan

Required by **SMARA**, a comprehensive plan for converting **mined lands** to a post-mining use, or **end use**, through **reclamation**. Every mining **operator** is required to obtain **lead agency** approval of such a plan.

Recontouring

Part of the **reclamation** process in which the filling, grading, and reshaping of mined lands occurs to prepare the mine site for its post-mining use, or **end use**.

Recycling

The crushing and processing of asphalt or concrete for re-use as **aggregate**, typically **base rock**.

Glossary

Regional water quality control board
Regional state agencies responsible for administering **California's Porter-Cologne Water Quality Control Act** and the federal **Clean Water Act**. Each board is responsible for administering these laws for their designated water basin. Each board must also adopt its own regional water quality control plan, frequently called a basin plan.

Regional water quality control plan
A plan adopted by a **regional water quality control board** that identifies **beneficial uses** within their basins and identifies water quality objectives to ensure protection of those uses and the prevention of nuisance.

Reserves
The portion of a site's **resources** that is considered economically mineable.

Resources
The body of demonstrated or inferred **minerals** existing at a particular site.

Responses to comments
Contained in the **final environmental impact** report, the **lead agency**'s written responses to all the comments received from the public and interested agencies during **circulation** of the **draft environmental impact report**.

Revegetation
Part of the **reclamation** process in which mined lands are planted with vegetation suitable for the mine site's **end use** and for controlling erosion.

Ringelmann test
A standard employed by many air pollution control districts to regulate dust and smoke emissions. The observer conducting the tests holds up a set of cards, ranging from white to black, in a line of site the distant emissions in the background. The observer then compares the color of the emissions to the color shown by the cards and assigns a number to the emission. Some districts do not allow dust or smoke emissions to exceed Ringelmann No. 1, corresponding to the card showing the lightest shade of gray.

RWB
Regional water quality control board

Scoping
The first stage in the preparation of an **environmental impact report**, in which the **lead agency** solicits comments on its scope, focus, and contents from other interested agencies and members of the public.

Screening
A condition **lead agencies** often impose to control the aesthetics of a mine site. The conditions may impose limitations on the size of stockpiles, mandate construction of earthen berms, or require planting of trees and vegetation.

Second use (end use)
The post-mining land use to which a mine site will be reclaimed following conclusion of mining.

Section 404
The section of the **Clean Water Act** that requires those who discharge dredged or fill material into **waters of the United States**, including **jurisdictional wetlands**, to obtain a permit from the **Army Corps of Engineers**.

Section 404 permit
A common reference to **dredge or fill permits** under Section 404 of the Clean Water Act.

Significant mineral resource
A designation the **State Mining and Geology Board** may give to areas classified **MRZ-2a** or **MRZ-2b**. The Board may designate an area significant from either a statewide or regional standpoint. The designations trigger the additional **lead agency** responsibility that the importance of affected mineral resources be considered and that mineral values be balanced against alternative land uses when considering any mining application.

SIP
State implementation plan

SMARA
Acronym for the Surface Mining and Reclamation Act of 1975, a comprehensive California law covered extensively in chapters 3–6.

SMGB
State Mining and Geology Board

Solution mining
Sometimes called **heap mining**, this method of **surface mining** uses chemicals to leech **metallic minerals** from previously excavated piles or "heaps." The leeched minerals are collected in liners underneath the heaps and transported into solution ponds where they are processed.

State Geologist
Head of the **California Geological Survey**, an agency within the **Department of Conservation**, responsible under SMARA for **mineral land classification**.

State implementation plan
Required by the **Clean Air Act**, a plan that states must adopt to provide for the implementation, maintenance and enforcement of the **National Ambient Air Quality Standards**.

State Mining and Geology Board
A nine-member board within the **Department of Conservation** that possesses policy-making authority under SMARA, which it exercises by promulgating regulations interpreting and implementing the Act. The Board also possesses certain enforcement powers and acts as an appellate body for lead agency decisions involving reclamation plans, financial assurances, and permits.

State Water Resources Control Board
An agency of the State of California that has adopted the **California Water Plan** and several policies to ensure protection of the state's water resources and water quality.

Statement of overriding considerations
Under **CEQA**, **findings** an agency may prepare to explain why it will approve a project even though one or more of the project's environmental impacts cannot be avoided or mitigated. The agency prepares such findings when it concludes that social, economic, legal, technical, or other benefits outweigh the project's significant environmental impacts.

Stockpile
A pile or storage location for certain materials. At a typical mine site, mineral products will be stockpiled in separate piles after they have been crushed and screened. There, the materials are stored until the operator receives an order for them, after which they are loaded onto a truck or barge for job-site delivery.

Stormwater Pollution Prevention Plan
A plan mining operations must have approved to control stormwater discharges from their operations.

Glossary

Streambed alteration agreement

An authorization required from the **California Department of Fish and Game** for any activity that would (1) substantially divert or obstruct the natural flow of, or (2) substantially change or use the material from, the bed, bank, or channel of a river, stream, or lake. *See* Cal. Fish & Game Code § 1602.

Strip mining

A term often used to describe **surface mining** generally, but that more accurately refers to **open-cast mining**.

Surface mining

The removal of minerals from openings in the earth's surface. Types of surface mining include **open-pit mining, open-cast mining, augur mining, dredging**, and **solution mining**. *See* Cal. Pub. Res. Code § 2735 (defining "surface mining operations" under **SMARA**).

Surface mining operation

Under **SMARA**, "[a]ll, or any part of, the process involved in the mining of minerals on mined land by removing overburden and mining directly form the mineral deposits, open-pit mining of minerals naturally exposed, mining by the augur method, dredging and quarrying, or surface work incident to an underground mine." Cal. Pub. Res. Code § 2735. Also includes "in place distillation or retorting or leaching"; "the production and disposal of mining waste"; and "prospecting and exploratory activities." *Id.*

Surface Mining Ordinance

An ordinance **SMARA** requires cities and counties to adopt whenever there is at least one **surface mining operation** within their jurisdiction or they receive their first application for a new mining operation. The ordinance establishes procedures for approval of **reclamation plans** and **financial assurances** and does not become effective until certified by the **State Mining and Geology Board**.

SWPPP

Stormwater Pollution Prevention Plan

Takeover of Lead Agency Authority

Under **SMARA**, the assumption by the **State Mining and Geology Board** of one or more **lead agency** enforcement powers. To takeover any power, the SMGB must find that the lead agency has (1) approved **reclamation plans** or **financial assurances** that do not comply with SMARA; (2) failed to conduct **annual inspections**; (3) intentionally misrepresented inspections; (4) failed to forfeit **financial assurances** when operators become **financially incapable of remaining**; (5) failed to take enforcement actions; or (6) failed to submit required information to the **Department of Conservation**. The SMGB must follow notice and hearing requirements before assuming any of these powers.

Taking (of endangered or threatened species)

An action prohibited under the **Endangered Species Act** as to **endangered** or **threatened species**. Under the Act, to take means to "harass, harm, pursue, hunt, shoot, wound, kill, trap, capture, or collect, or to attempt to engage in any such conduct." 16 U.S.C. § 1532(19).

Tarping

A requirement that trucks **hauling** materials cover their trailers to prevent dust and debris from falling onto roads. Under the California Vehicle Code, truckers may not be forced to cover loads of aggregate. *See* Cal. Veh. Code § 23114(e)(4).

Threatened species

Under the **Endangered Species Act**, any species that "is likely to become an endangered species within the foreseeable future throughout all or a significant portion of its range.... 16 U.S.C. § 1532(20).

Track out

Rocks, gravel, sand, mud, dust, or debris tracked outside a mine site by trucks **hauling** material.

Underground mining

The removal of minerals from beneath the earth's surface.

United States Fish and Wildlife Service

The federal agency responsible for administering the **Endangered Species Act**. The agency's responsibilities include listing endangered and threatened species and issuing incidental take permits under the Act.

Washing

The use of water in processing operations to remove fine and clayish particles from excavated materials. Washing is necessary in asphalt production since the presence of fines and particles inhibits bonding of the petroleum agents that are mixed with crushed rock.

Waste discharge requirements

Under the **Porter-Cologne Water Quality Control Act**, an authorization dischargers must obtain from a **regional water quality control board** prior to discharging any waste into waters of the state.

Waste discharge to land

Under the **Porter-Cologne Water Quality Control Act**, the indirect discharge into waters of the state from land disposal of mining wastes.

Waste pile

Uneconomic material separated from the mined commodity and stored, usually for **recontouring** during **reclamation**.

Waste water

Water left over from washing operations that must be disposed of in accordance with **waste discharge requirements**.

Water quality criteria

Under the **Clean Water Act**, criteria states must adopt to protect the **beneficial uses** designated in their water quality standards. The criteria must "be based on sound scientific rationale and must contain sufficient parameters or constituents to protect the designated use." 40 C.F.R. § 131.11(a).

Water quality standards

Required by the **Clean Water Act**, standards every state must adopt in specifying the **beneficial uses** to be achieved and protected. The standards must "take into consideration the use and value of water for public water supplies, protection and propagation of fish, shellfish, and wildlife, recreation in and on the water, agricultural, industrial, and other purposes including navigation." 40 C.F.R. § 131.10(a).

Waters of the United States

The waters regulated by the **Clean Water Act**, which includes navigable waters and **wetlands**. Because the extent of that Act's regulation is bound by the Commerce Clause of the United States Constitution, purely intra-state waters are not covered. *See Rapanos v. United States*, 126 S. Ct. 2208 (2006); *Solid Waste Agency of N. Cook County v. United States Army Corps of Engineers*, 531 U.S. 159 (2001).

Glossary

WDRs

Permit issued by a RWQCB for discharge of waste into waters of the state. A Report of Waste Discharge must be filed with the appropriate RWQCB.

Wet scrubber

An air pollution abatement device that utilizes high energy water or reactant spray to remove pollutants, either by absorption or reaction.

Wetland

Under the **Clean Water Act**, an area with soils that (1) are periodically or permanently saturated or inundated; (2) are hydric, exhibiting anaerobic conditions in their upper parts; and (3) contain hydrophytic vegetation plants that grow where little oxygen is present. 33 C.F.R. § 328.3(b).

Wetlands delineation

A study performed by private consultant to determine the existence and extent of **jurisdictional wetlands** located on a property or potential mine site. The study may be certified by the **United States Army Corps of Engineers** for a period of three years.

Williamson Act Contract

A contract between a land owner and city or county in which the owner agrees to use its property only for agricultural or open space uses in exchange for preferential property tax treatment. In some cases, mine property may be placed in a Williamson Act contract. *See* Cal. Gov't Code §§ 51238.1, 51238.2, 51238.3.

Working area

The area of a **mine** or **quarry** where excavation of minerals occurs.

Working face

The specific side or "face" of a **pit** or **working area** where mineral extraction takes place.

Zoning

The regulation of the use of land, the size of buildings, and the developable areas of legal parcels. Zoning is accomplished through the classification and placement of land uses in discrete districts or zones.

Zoning district

A class of land uses for purposes of **zoning**. Only uses that are similar to or compatible with each other are placed in the same district; other uses are excluded to avoid land use conflicts and the creation of nuisances. Typical districts include residential, commercial, industrial, agricultural, and open space.

Zoning ordinance

A city or county's legislative enactment of **zoning**. The zoning ordinance comprehensively establishes all **zoning districts** and sets forth the requirements for each district.

List of Acronyms

- ARA = Aggregate resource area
- BACT = Best available control technology
- BMP = Best management practice
- CAA = Clean Air Act
- CARB = California Air Resources Board
- CEQA = California Environmental Quality Act
- CESA = California Endangered Species Act
- CGS = California Geological Survey
- Corps = U.S. Army Corps of Engineers
- CUP = Conditional use permit
- CWA = Clean Water Act
- DFG = Department of Fish and Game
- EIR = Environmental impact report
- EPA = Environmental Protection Agency
- ESA = Endangered Species Act
- HCP = Habitat conservation plan
- LAER = Lowest available emission rate
- NAAQS = National Ambient Air Quality Standards
- NIMBY = Not in My Backyard
- NPDES = National Pollutant Discharge Elimination System
- NSR = New source review
- NWP = Nationwide Permit Program
- OMR = Office of Mine Reclamation
- OPR = Governor's Office of Planning and Research
- PM10 = Particulate matter
- PSD = Prevention of significant deterioration
- RWB = Regional water quality control board
- SIP = State Implementation Plan
- SMARA = Surface Mining and Reclamation Act
- SMGB = State Mining and Geology Board
- SWB = State Water Resources Control Board
- SWPPP = Storm water pollution prevention plan
- USFWS = U.S. Fish and Wildlife Service
- WDR = Waste discharge requirement

Table of Authorities

CASES

ATTORNEY GENERAL OPINIONS

UNITED STATES CONSTITUTION

FEDERAL STATUTES

FEDERAL REGULATIONS

OTHER FEDERAL SOURCES

CALIFORNIA STATUTES

Code of Civil Procedure

Fish and Game Code

Government Code

Health and Safety Code

Public Contract Code

Public Resources Code

Vehicle Code

Water Code

CALIFORNIA CODE OF REGULATIONS

(for CEQA Guidelines, Title 14, *see* next section)

Title 13

Title 14

CEQA GUIDELINES

CALIFORNIA LEGISLATIVE HISTORY MATERIAL

OTHER CALIFORNIA SOURCES

AIR QUALITY MANAGEMENT DISTRICT RULES AND REGULATIONS

ARTICLES AND PUBLICATIONS

Index

B

C

D

G

H

I

J

K

L

N

Q

R

S

T

U

V

Other Guides and References

PLANNING . LAND USE . URBAN AFFAIRS . ENVIRONMENTAL ANALYSIS . REAL ESTATE DEVELOPMENT

Curtin's California Land Use and Planning Law

Well-known, heavily quoted, definitive summary of California's planning laws with expert commentary on the latest statutes and case law. Includes practice tips, graphics, a table of authorities, and an index. Cited by the California Courts, including the California Supreme Court, as an Authoritative Source.

Cecily T. Talbert • Revised annually

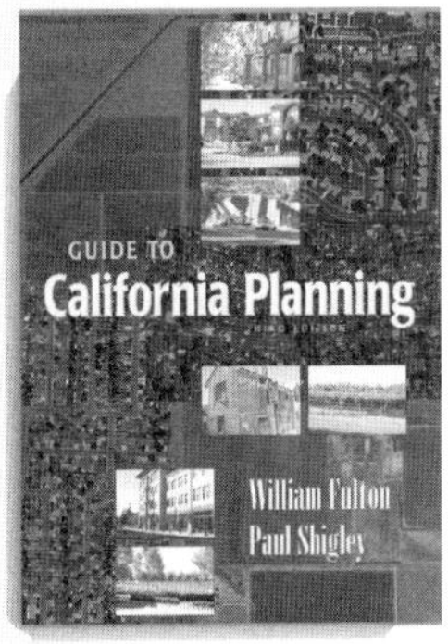

Guide to California Planning

All-new edition describes how planning really works in California, how cities, counties, developers, and citizen groups all interact with each other to shape California communities and the California landscape, for better and for worse. Recipient of the California Chapter APA Award for Planning Education.

William Fulton and Paul Shigley
2005 (third) edition

California Environmental Law and Policy: A Practical Guide

The only book that covers the entire field of California environmental, land use, and natural resources law in a concise, user-friendly format. Comprehensively surveys the most important federal and state environmental statutes and regulatory programs. Highlights key permits, landmark court cases, and current policy issues. Provides practical tips and tools to guide your project through the regulatory process successfully.

Herson and Lucks • 2008

Ballot Box Navigator: A Practical and Tactical Guide to Land Use Initiatives and Referenda in California

The authoritative resource on securing a ballot title, qualifying an initiative or referendum for the ballot, and submitting a measure for an election. With short articles, practice tips, drawings, an index, glossary, and a table of authorities.

Michael Patrick Durkee et al. • 2003

California Public Contract Law: Basic Principles and Special Requirements

Public Contract Law is an easy-to-use, concise reference on State of California laws governing the public contracting processes used by federal, state, regional, and local agencies in California. The book reviews key requirements, suggests model provisions, and provides extensive references for relevant statutes and regulations. An important resource for bidders, agency administrators, and contractors involved in public contracting law in California.

Jeremy G. March • 2007

California School Facilities Planning: A Guide to Laws and Procedures for Funding, Siting, Design, and Construction

A single-source reference that offers a thorough discussion of laws and regulations that govern planning, funding, siting, design, and construction of educational facilities. The book will guide the reader through every stage of the planning process—from initial conception through construction.

Maureen F. Gorsen et al. • 2005

CALL TOLL-FREE
(800) 931-9373 OR FAX (707) 884-4109

Solano Press Books

www.solano.com . spbooks@solano.com . facsimile 707 884-4109

Solano Press Books

www.solano.com

spbooks@solano.com

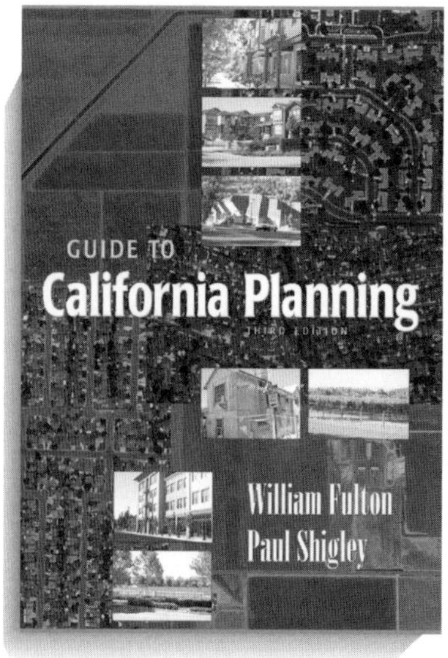

Guide to California Planning

by William Fulton and Paul Shigley

Since the book was first published in 1991, *Guide to California Planning* has served as the authoritative textbook on city and county planning practice. It is used in virtually every college- and graduate-level planning program in California.

In this revised, expanded third edition, the authors lay out planning laws and processes in detail, describing how planning really works—how cities and counties and developers and citizen groups all interact on a daily basis to shape California communities and the California landscape, for better and for worse. Easy to read and understand, *Guide to California Planning* is far more than a textbook. It's also an ideal tool for members of allied professions in the planning and development fields as well as citizen activists.

The Planning Commissioner and the California Dream

Plan It Again, Sam

by Marjorie W. Macris, FAICP

Tackling a serious subject with whimsy and humor, as well as seriousness of purpose, the book offers specific guidelines on how to function as an effective planning commissioner in the face of complex issues, special interests, committed advocates, and intricate planning and environmental laws. Describing their roles and responsibilities, and how to make sure things go smoothly, *The Planning Commissioner and the California Dream* includes interviews with a variety of commissioners, a glossary of common terms, and tips on how to do the job well. Peppered with amusing references to selected motion pictures that convey the significance of California and its unique environment.

Solano Press Books

www.solano.com

spbooks@solano.com

California Public Contract Law

Basic Principles and Special Requirements

by Jeremy G. March

California Public Contract Law is an easy-to-use, concise reference on the laws governing the public contracting processes used by federal, state, regional, and local agencies in California. The book reviews key requirements, suggests model provisions, and provides extensive references for relevant statutes and regulations. An important resource for agency administrators, contractors, and bidders involved in public contracting law.

Includes a list of useful legal research sites on the Internet, a summary of frequently used provisions of the Public Utilities Code and the Public Contract Code, as well as a table of authorities, a list of acronyms, and an index.

Telecommunications

The Governmental Role in Managing the Connected Community

by Paul Valle-Riestra

Telecommunications explains this dynamic field in straightforward language useful to planners, attorneys, local legislators, cable franchise administrators, and company personnel. The book offers a detailed analysis of federal and state laws governing the location and regulation of physical facilities including cable, traditional telephone systems, wireless systems (cellular, paging, and Internet), satellite dishes, and antennas.

Public policy associated with telecommunications is discussed, along with issues of siting and rights-of-way, current technologies, the future of regulation, and environmental and design standards. Includes practice tips, charts, photographs, a glossary, table of authorities, and an index.

California Water II

by Arthur L. Littleworth
and Eric L. Garner

This all-new second edition of *California Water* is the first book since Wells Hutchins' 1956 "bible," *California Law of Water Rights,* to serve as a comprehensive guide to historical, legal, and policy issues that affect the use of water in California.

The book is a major resource for local officials—in water districts, cities, and counties—as well as for lawyers, judges, engineers, planners, community leaders, environmentalists, developers, and farmers. *California Water II* addresses the question, "Do we have enough water?" and covers key developments in water law: the implementation of the historic Bay-Delta Accord, the crisis in the Delta, settlement of critical issues on the Colorado River, global warming, and the emergence of new water supplies such as water transfers, conservation, recycling, and the conjunctive use of groundwater.

Water and Land Use
Planning Wisely for California's Future

by Karen E. Johnson
and Jeff Loux

This is the first complete guide to address the increasingly important link between land use planning in California and the availability of water. *Water and Land Use* summarizes key statutes, governmental policies and requirements, and current practices. In addition, the book presents methods for evaluating water demand, supply, reliability, and quality, and for meeting legislative requirements for linking water supplies and land use decisions.

A valuable "how to" handbook for environmental and land use planners, engineers, water resource planners, government officials, and attorneys. With illustrations and photographs, tables, flow charts, case studies, sample documents, practice tips, a glossary, references, and an index.

PLANNING

LAND USE

URBAN AFFAIRS

ENVIRONMENTAL ANALYSIS

REAL ESTATE DEVELOPMENT

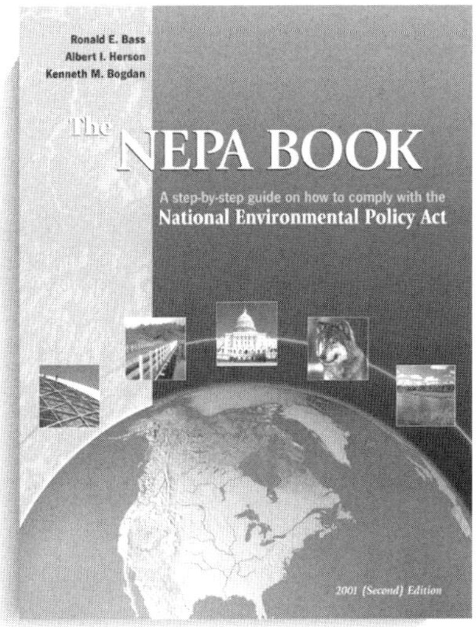

The NEPA Book

A Step-by-Step Guide on How to Comply with the National Environmental Policy Act

by Ronald E. Bass, Albert I. Herson, and Kenneth M. Bogdan

A practitioner's handbook that takes you through the basic requirements and most important decision points of the National Environmental Policy Act. The book will help you determine if a particular action is subject to NEPA, decide whether an environmental assessment or a FONSI is appropriate, prepare legally-adequate environmental documents, write effective mitigation measures, and understand the relationship between NEPA and other environmental laws.

Ideal for environmental specialists, agency reviewers and decisionmakers, attorneys, consultants, students, and citizens who want to effectively influence the environmental decisions that shape their communities.

Guide to the California Forest Practice Act and Related Laws

by Sharon E. Duggan and Tara Mueller

A comprehensive treatise on legislation that regulates timber harvesting on private lands in California, covering state and federal statutory and regulatory requirements, case law, and agency policies. *Guide to the California Forest Practice Act* includes an in-depth discussion of the California Environmental Quality Act; regulation of pesticides; restrictions on streambed alteration; timber harvesting in the coastal zone; common law nuisance and the public trust doctrine; and protection of water quality, endangered and threatened species, and wild and scenic rivers.

Intended as a complete resource for the full range of actors involved, the book includes case studies, short articles, charts, graphs, tables, photos, appendices, a table of authorities, and an index.

PLANNING

LAND USE

URBAN AFFAIRS

ENVIRONMENTAL ANALYSIS

REAL ESTATE DEVELOPMENT

CALL TOLL-FREE
(800) 931-9373
OR FAX (707) 884-4109

Solano Press Books

www.solano.com

spbooks@solano.com

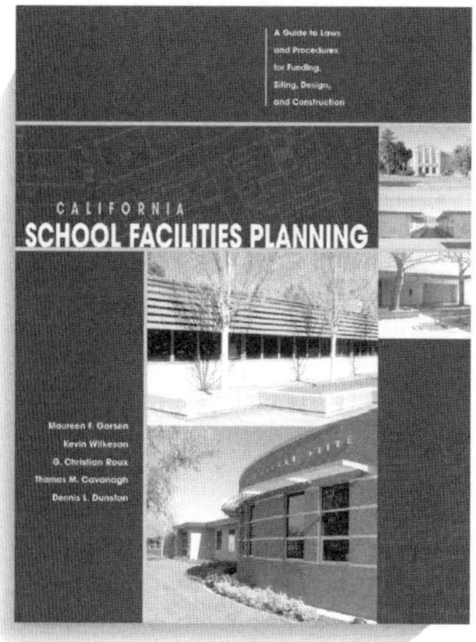

California School Facilities Planning

A Guide to Laws and Procedures for Funding, Siting, Design, and Construction

by Maureen F. Gorsen, Kevin Wilkeson, G. Christian Roux, Thomas M. Cavanagh, and Dennis L. Dunston

A single-source reference that offers a thorough discussion of laws and regulations that govern planning, funding, siting, design, and construction of educational facilities. *School Facilities Planning* will guide the reader chronologically through every stage of the planning process—from initial conception through construction.

Topics covered include the history of school facilities nationwide; funding sources and financing methods; planning and design, both for individual schools and district-wide; and school siting issues, including acquisition of property and compliance with land use and environmental laws.

Planning for Child Care in California

by Kristen M. Anderson

Presents basic information and guidelines for municipal, county, and school district planners, and for child care professionals and their advocates. It features strategies for ensuring that child care needs are met locally, with resource material that explains how it is regulated in its various forms through the General Plan and zoning.

Planning for Child Care in California discusses guidelines for incorporating child care goals into the public planning process, with specific attention to the location, planning, and design of housing, centers of employment, and transit-based facilities. Numerous examples of real child care projects, designs, and partnerships are included, along with sources of funding and other implementation strategies.